Lecture Notes in Physics

For information about Vols. 1–110, please contact your bookseller or Springer-Verlag.

Vol. 111: H.-J. Schmidt, Axiomatic Characterization of Physical Geometry. V, 163 pages. 1979.

Vol. 112: Imaging Processes and Coherence in Physics. Proceedings, 1979. Edited by M. Schlenker et al. XIX, 577 pages. 1980.

Vol. 113: Recent Advances in the Quantum Theory of Polymers. Proceedings 1979. Edited by J.-M. André et al. V, 306 pages. 1980.

Vol. 114: Stellar Turbulence. Proceedings, 1979. Edited by D. F. Gray and J. L. Linsky. IX, 308 pages. 1980.

Vol. 115: Modern Trends in the Theory of Condensed Matter. Proceedings, 1979. Edited by A. Pekalski and J. A. Przystawa. IX, 597 pages. 1980.

Vol. 116: Mathematical Problems in Theoretical Physics. Proceedings, 1979. Edited by K. Osterwalder. VIII, 412 pages. 1980.

Vol. 117: Deep-Inelastic and Fusion Reactions with Heavy Ions. Proceedings, 1979. Edited by W. von Oertzen. XIII, 394 pages. 1980.

Vol. 118: Quantum Chromodynamics. Proceedings, 1979. Edited by J. L. Alonso and R. Tarrach. IX, 424 pages. 1980.

Vol. 119: Nuclear Spectroscopy. Proceedings, 1979. Edited by G. F. Bertsch and D. Kurath. VII, 250 pages. 1980.

Vol. 120: Nonlinear Evolution Equations and Dynamical Systems. Proceedings, 1979. Edited by M. Boiti, F. Pempinelli and G. Soliani. VI, 368 pages. 1980.

Vol. 121: F. W. Wiegel, Fluid Flow Through Porous Macromolecular Systems. V, 102 pages. 1980.

Vol. 122: New Developments in Semiconductor Physics. Proceedings, 1979. Edited by F. Beleznay et al. V, 276 pages. 1980.

Vol. 123: D. H. Mayer, The Ruelle-Araki Transfer Operator in Classical Statistical Mechanics. VIII, 154 pages. 1980.

Vol. 124: Gravitational Radiation, Collapsed Objects and Exact Solutions. Proceedings, 1979. Edited by C. Edwards. VI, 487 pages. 1980.

Vol. 125: Nonradial and Nonlinear Stellar Pulsation. Proceedings, 1980. Edited by H. A. Hill and W. A. Dziembowski. VIII, 497 pages. 1980.

Vol. 126: Complex Analysis, Microlocal Calculus and Relativistic Quantum Theory. Proceedings, 1979. Edited by D. Iagolnitzer. VIII, 502 pages. 1980.

Vol. 127: E. Sanchez-Palencia, Non-Homogeneous Media and Vibration Theory. IX, 398 pages. 1980.

Vol. 128: Neutron Spin Echo. Proceedings, 1979. Edited by F. Mezei. VI, 253 pages. 1980.

Vol. 129: Geometrical and Topological Methods in Gauge Theories. Proceedings, 1979. Edited by J. Harnad and S. Shnider. VIII, 155 pages. 1980.

Vol. 130: Mathematical Methods and Applications of Scattering Theory. Proceedings, 1979. Edited by J. A. DeSanto, A. W. Sáenz and W. W. Zachary. XIII, 331 pages. 1980.

Vol. 131: H. C. Fogedby, Theoretical Aspects of Mainly Low Dimensional Magnetic Systems. XI, 163 pages. 1980.

Vol. 132: Systems Far from Equilibrium. Proceedings, 1980. Edited by L. Garrido. XV, 403 pages. 1980.

Vol. 133: Narrow Gap Semiconductors Physics and Applications. Proceedings, 1979. Edited by W. Zawadzki. X, 572 pages. 1980.

Vol. 134: $\gamma\gamma$ Collisions. Proceedings, 1980. Edited by G. Cochard and P. Kessler. XIII, 400 pages. 1980.

Vol. 135: Group Theoretical Methods in Physics. Proceedings, 1980. Edited by K. B. Wolf. XXVI, 629 pages. 1980.

Vol. 136: The Role of Coherent Structures in Modelling Turbulence and Mixing. Proceedings 1980. Edited by J. Jimenez. XIII, 393 pages. 1981.

Vol. 137: From Collective States to Quarks in Nuclei. Edited by H. Arenhövel and A. M. Saruis. VII, 414 pages. 1981.

Vol. 138: The Many-Body Problem. Proceedings 1980. Edited by R. Guardiola and J. Ros. V, 374 pages. 1981.

Vol. 139: H. D. Doebner, Differential Geometric Methods in Mathematical Physics. Proceedings 1981. VII, 329 pages. 1981.

Vol. 140: P. Kramer, M. Saraceno, Geometry of the Time-Dependent Variational Principle in Quantum Mechanics. IV, 98 pages. 1981.

Vol. 141: Seventh International Conference on Numerical Methods in Fluid Dynamics. Proceedings. Edited by W. C. Reynolds and R. W. MacCormack. VIII, 485 pages. 1981.

Vol. 142: Recent Progress in Many-Body Theories. Proceedings. Edited by J. G. Zabolitzky, M. de Llano, M. Fortes and J. W. Clark. VIII, 479 pages. 1981.

Vol. 143: Present Status and Aims of Quantum Electrodynamics. Proceedings, 1980. Edited by G. Gräff, E. Klempt and G. Werth. VI, 302 pages. 1981.

Vol. 144: Topics in Nuclear Physics I. A Comprehensive Review of Recent Developments. Edited by T.T.S. Kuo and S.S.M. Wong. XX, 567 pages. 1981.

Vol. 145: Topics in Nuclear Physics II. A Comprehensive Review of Recent Developments. Proceedings 1980/81. Edited by T. T. S. Kuo and S. S. M. Wong. VIII, 571-1.082 pages. 1981.

Vol. 146: B. J. West, On the Simpler Aspects of Nonlinear Fluctuating. Deep Gravity Waves. VI, 341 pages. 1981.

Vol. 147: J. Messer, Temperature Dependent Thomas-Fermi Theory. IX, 131 pages. 1981.

Vol. 148: Advances in Fluid Mechanics. Proceedings, 1980. Edited by E. Krause. VII, 361 pages. 1981.

Vol. 149: Disordered Systems and Localization. Proceedings, 1981. Edited by C. Castellani, C. Castro, and L. Peliti. XII, 308 pages. 1981.

Vol. 150: N. Straumann, Allgemeine Relativitätstheorie und relativistische Astrophysik. VII, 418 Seiten. 1981.

Vol. 151: Integrable Quantum Field Theory. Proceedings, 1981. Edited by J. Hietarinta and C. Montonen. V, 251 pages. 1982.

Lecture Notes in Physics

190

Karl Kraus

States, Effects, and Operations

Fundamental Notions of Quantum Theory

Lectures in Mathematical Physics
at the University of Texas at Austin

Edited by A. Böhm, J. D. Dollard and W. H. Wootters

Springer-Verlag
Berlin Heidelberg GmbH 1983

Author

Karl Kraus
Physikalisches Institut der Universität Würzburg
Am Hubland, D-8700 Würzburg

Editors

A. Böhm
Physics Department, University of Texas
Austin, TX 78712, USA

J. D. Dollard
Mathematics Department, University of Texas
Austin, TX 78712, USA

W. H. Wootters
Department of Physics and Astronomy, Williams College
Williamstown, Mass. 01267, USA

ISBN 978-3-540-12732-1

Library of Congress Cataloging in Publication Data. Kraus, Karl, 1938- States effects, and operations. (Lecture notes in physics; 190) 1. Quantum theory. I. Title. II. Series. QC174.12.K72 1983 530.1'2 83-16906

ISBN 978-3-540-12732-1 ISBN 978-3-540-38725-1 (eBook)
DOI 10.1007/978-3-540-38725-1

Originally published by Springer-Verlag Berlin Heidelberg New York Tokyo in 1983

2153/3140-543210

Dem Andenken meiner Mutter

Marie Kraus geb. Paus

4. Juli 1915 - 13. Oktober 1980

Preface

The lecture notes of this volume are based on a series of lectures given in the mathematical physics program at the University of Texas at Austin. These lecture series were started in 1976 with the purpose of establishing communication between mathematicians and physicists and of informing graduate students of both departments about recent developments in mathematics and physics. The lectures are directed at "non-specialists", and this volume also should be suitable for a general audience.

The subject of this volume is the interpretation and the foundations of quantum theory. There exist diverging opinions about this subject. Some people believe that quantum physics is so remote from the usual way of thinking and so foreign to everyday language that its meaning cannot be communicated directly. According to them, the only way to an understanding of quantum mechanics is to study many separate problems and thereby acquire an intuitive feeling for the foundations of quantum mechanics. The premise for the work discussed here is that the foundations of quantum mechanics <u>can</u> be verbalized, and the present lectures are an attempt to communicate them.

But the champions of this field also have different points of view concerning the interpretation of quantum mechanics. The purpose of these lecture notes is not to give a comparative review of the different approaches - this can be found in "Interpretations and Foundations of Quantum Theory", H. Neumann (Ed.), BI-Wissenschaftsverlag (1981). Here just one approach is presented, the one developed by Ludwig and his school, which appears to be sufficiently general to encompass most of the other orthodox (Copenhagen) interpretations of quantum mechanics.

The usual custom in this field is to start from physically meaningful postulates in a very general mathematical language, and derive from them the usual formalism of quantum mechanics in terms of linear operators in linear scalar product spaces. Even restricting oneself to only one approach, this is still a very tedious procedure. A more accessible road to the foundations of quantum mechanics is to start with the mathematically more familiar objects of operators and vectors in a Hilbert space, rather than deriving them, and to formulate the fundamental statements in terms of these mathematical quantities. This is the road that has been followed in these lectures in order to make this important and fundamental subject accessible to a larger audience.

Arno Böhm
J.D. Dollard
W.H. Wootters
(Editors)

Author's Preface

The subject of these notes is ordinary quantum mechanics. More specifically, they deal with some particular aspects of the conceptual structure of that theory, which - at least in this form and to this extent - are not treated in the usual textbooks. The reader I had in mind when writing these notes, therefore, should already be familiar with the standard formulation of quantum mechanics, and should be interested not only in applications to specific physical problems but also in the general structure and the physical interpretation of the theory.

As quantum mechanics is usually formulated in terms of vectors and linear operators in Hilbert space, some basic knowledge of this mathematical framework has been assumed. Whenever more advanced mathematical tools are required, I have tried to explain them - including suitable references, if necessary - in order to facilitate the reading. For the same reason, most mathematical deductions are presented in full detail. I hope that readers who are not too much interested in such technicalities will just glance over them, and concentrate their attention on the more essential statements.

These notes are based on a series of lectures given at the University of Texas at Austin in September 1980. An earlier account of the topics discussed in these lectures has been published in the proceedings of the conference on "Foundations of Quantum Mechanics and Ordered Linear Spaces" at Marburg in 1973 [1]. The invitation to lecture on these matters at the University of Texas at Austin provided me with an occasion to rethink the whole subject, to extend my results in several aspects, and to reformulate everything in a more selfcontained way.

For this invitation, and for their kind hospitality during my pleasant stay at Austin, I want to thank Profs. A. Bohm, J. Dollard, T.A. Griffy, and J.A. Wheeler. I am also very grateful to Dr. W.H. Wootters (Austin), who read a preliminary version of my notes, corrected my English, and suggested several other improvements. Last but not least, I want to express my hearty thanks to Mrs. Ch. Steinbauer and Dr. W. Petzold (Würzburg), who have been burdened with the tedious task of typewrinting and proofreading, and to Prof. W. Beiglböck from Springer Verlag for his great patience and constant encouragement during the preparation of the final typescript.

Contents

§1 States and Effects

Almost all treatments of quantum mechanics agree in ascribing fundamental importance to the notions of "state" and "observable". The physical interpretation of these notions, however, differs considerably from one author to another. The discussion of as many as possible of these different points of view is not the subject of the present investigation, in view of both the limited space available and the insufficient familiarity of the author with most of them.

We prefer instead to follow here as closely as possible one particular interpretation of quantum mechanics, which has been elaborated by Ludwig and his school [2], and which the present author considers to be particularly satisfactory. A detailed exposition of Ludwig's theory would still go far beyond the limits of space available here. We shall try to sketch some of the main ideas only, and to encourage any reader who is interested in more details to consult the original literature quoted above.

Another reason for not entering a detailed discussion of various interpretations of quantum mechanics is our conviction that the more technical parts of the present work, if suitably reformulated, are compatible with most or even all of them. We thus believe that even a reader who disagrees completely with the interpretation proposed here many profit from the following investigations, even if the burden of reinterpreting our results in terms of his preferred interpretation of quantum mechanics is entirely left to himself.

According to Ludwig, any physical theory is in some sense to be interpreted "from outside", i.e., in terms of "pretheories" not belonging to the theory in question itself. For quantum mechanics in particular, these "pretheories" belong to the realm of classical physics, and describe the construction and application of macroscopic preparing and measuring instruments. A similar point of view has been advocated by Bohr, who repreatedly stressed the importance of the classical nature of measuring instruments for the understanding of quantum mechanics.

Such statements should not be misinterpreted to mean that the behavior of macroscopic instruments can always be understood completely in terms of classical theories such as, e.g., mechanics and electrodynamics. If this were true, quantum mechanics would never have been invented. Indeed, quantum mechanics just serves to describe some particular behavior of macroscopic instruments which can not be explained classically.

It is maintained, however, that the construction and application of instruments can be - and, in practice, always are - described in purely classical terms, without any reference to quantum mechanics. Moreover, in the same spirit, typical changes occuring in such instruments during "measurements", such as, e.g., the discharge of a counter, are accepted as objectively real events, in much the same "naive" way in which experimentalists always accept such occurrences in practice, and as we all do in everyday life. The physical interpretation of quantum mechanics is then formulated entirely in terms of such instruments and events, and thus ultimately rests on ascribing "objective reality" to (at least) the macroscopic would surrounding us. In this respect the present interpretation of quantum mechanics profoundly disagrees with certain other interpretations, ascribing, e.g., a descisive rôle to human consciousness in the creation of observable events. It is in fact one of the main motivations of Ludwig's theory to show that quantum mechanics, with all its subtleties, can be formulated and interpreted consistently without such radical changes in our everyday concept of physical reality. (Nevertheless, this "naive" concept has to be refined considerably to cover, finally, such "things" as atoms or electrons. This problem has also been analyzed by Ludwig, but the results of this deep analysis cannot even be sketched here.)

According to the point of view adopted here, the fundamental notions of quantum mechanics have thus to be defined operationally in terms of macroscopic instruments and prescriptions for their application. A preliminary notion of "state" is then most simply given in terms of preparing instruments. Experience tells us that suitably constructed instruments can be used to produce ensembles - in principle arbitrarily large - of single microsystems of the particular type considered (e.g., electrons). Ascribing something like a "state" to such an ensemble is then, at this lowest level of the theory, just a shorthand notation for the applied preparation procedure. We introduce, for this purpose, the notion of a "prestate". A prestate is thus specified by the technical description of the preparing instrument and its mode of application. Such a specification is abbreviated here by using labels w for prestates; accordingly, the same label w shall also be used to denote the applied preparing instrument - or rather, the entire preparation procedure - itself. Two ensembles of microsystems are thus in different prestates, $w_1 \neq w_2$, if and only if they are produced by different preparing procedures.

Another empirical fact is the existence of so called measuring instruments, which are capable of undergoing macroscopically observable changes due to ("triggered by") their interaction with single microsystems. The simplest type of measuring instrument is one on which just a single change may be triggered. For instance, an originally charged counter may be found either still charged or discharged, after it has been exposed to an electron emitted by some preparing apparatus. (The result will depend, loosely speaking, on the efficiency of the counter, and on whether or not the electron "hits" it.) Instruments of this type perform so called yes-no measurements: calling the observable change of the instrument an "effect", one usually defines the result of a single measurement to be "yes" if the effect occurs, and "no" if the effect does not occur. It is equally possible, however, and sometimes even appropriate, to associate "yes" with the non-occurrence of the effect, and vice versa. (With its reading reinterpreted in this way, the apparatus then performs a different - although closely related - yes-no measurement.)

For many purposes it is more convenient to associate "measured values" 1 and 0 with the results "yes" and "no", respectively. With this convention, yes-no measurements fit into a broader class of measurements, involving also instruments with, e.g., a movable pointer on a scale, which in general contains more than only two possible measured values. However, such more complicated instruments - and the general notion of "observables" connected with them - need not be considered when we first discuss the basic facts of quantum mechanics. It is indeed well known that the measurement of any observable can be interpreted, in a standard way, as a combination of yes-no measurements. The latter are thus not only simple prototypes of measurements, but also the elementary building blocks for more general ones. We shall return to this point in §6.

An instrument performing yes-no measurements is called an effect apparatus, and shall also be symbolized here by some letter, usually f. As in the case of preparing instruments, this label f stands for a complete technical description of the apparatus, including the instructions for its application and its reading.

Assume now a preparing instrument w produces a single microsystem, which then interacts with an effect apparatus f, leading in turn either to the occurrence or the non-occurrence of the corresponding effect on the apparatus f. Call this a "single experiment", and assume such single experiments, with given w and f, to be repeated N times. (Keeping w and f fixed means, of course, to use the same or at least identically constructed instruments in all single experiments.)

We may then also say that the effect apparatus f has been applied to an ensemble of N microsystems in the prestate w.

The effect apparatus f will yield the answer "yes" in N_+ single experiments, and "no" in the remaining $N_- = N - N_+$ cases. In general, N_+ will be neither N nor zero; i.e., the outcome of a single experiment will not be determined completely by the instruments w and f. Nevertheless, in each series of N single experiments with given w and f the fraction N_+/N comes out roughly the same, if only N is sufficiently large, such that N_+/N approaches a definite limit $\tilde{\mu}(f,w)$ for very large N. Thus experience tells us that effect apparatuses f are triggered with reproducible relative frequencies $\tilde{\mu}(f,w)$ by microsystems prepared in a prestate w. Deviations of the observed values of N_+/N from $\tilde{\mu}(f,w)$ can then be interpreted as statistical errors due to the finiteness of N.

This most crucial fact is the empirical basis for the statistical laws of quantum mechanics. In accordance with the usual terminology, we call $\tilde{\mu}(f,w)$ the "probability" for the triggering of the effect apparatus f in the prestate w. Whenever the notion of "probability" occurs in the present interpretation of quantum mechanics, it has to be understood as synonymous with "relative frequency", as in the particular case of $\tilde{\mu}(f,w)$ just considered. Since N_+/N is also the average of the measured values 1 (yes) and 0 (no) obtained in N single experiments, we may also call $\tilde{\mu}(f,w)$ the expectation value of the f measurement in the prestate w.

Every experimentalist knows that some minor technical details of a preparing instrument w or an effect apparatus f may be changed without affecting their statistical behavior, as expressed by the probability function $\tilde{\mu}(f,w)$. But even two completely different preparing intruments w_1 and w_2 may give rise to the same probabilities for the triggering of arbitrary effect apparatuses f; i.e.,

$$\tilde{\mu}(f,w_1) = \tilde{\mu}(f,w_2) \text{ for all } f. \tag{1.1}$$

Likewise, there certainly exist (slightly, or even completely, different) effect apparatuses f_1 and f_2, such that

$$\tilde{\mu}(f_1,w) = \tilde{\mu}(f_2,w) \text{ for all } w; \tag{1.2}$$

i.e., f_1 and f_2 are triggered with equal probabilities in arbitrary prestates w. If, as usual, we restrict our attention to the probabilities $\tilde{\mu}(f,w)$, as the basic quantities for the formulation of the statistical laws of quantum mechanics, then any differences between two preparing instruments or effect apparatuses which do not affect these probabilities become inessential. Accordingly, two prestates w_1 and w_2 satisfying (1.1), as well as two effect apparatuses f_1 and f_2 satisfying (1.2), are defined to be equivalent. If thus read as equivalence relations, Eqs. (1.1) and (1.2) define equivalence classes W and F of preparing instruments (prestates) w and effect apparatuses f, respectively. All instruments w or f in a given equivalence class W or F thus behave "statistically", with respect to the probabilities $\tilde{\mu}(f,w)$, in the same way. An equivalence class W is called a <u>state</u>, as usual, whereas - in accordance with Ludwig's terminology - we call an equivalence class F an <u>effect</u>. An ensemble of N microsystems, prepared by an instrument w in the equivalence class W, is called an ensemble in the state W, whereas an effect apparatus f in the equivalence class F is said to measure the effect F.

By definition of these equivalence classes, the function $\tilde{\mu}(f,w)$ gives rise to a new function $\mu(F,W)$, called the probability for the occurrence of the effect F in the state W, as defined by

$$\mu(F,W) = \tilde{\mu}(f,w) \tag{1.3}$$

with $f \in F$ and $w \in W$. By (1.1) and (1.2), this definition does not depend on the choice of particular representatives f of F and w of W on the right hand side of (1.3). Denoting the sets of states W and effects F by K and L, respectively, (1.3) thus defines a function μ on $L \times K$. According to its physical meaning, this function obviously satisfies

$$0 \leq \mu(F,W) \leq 1 \tag{1.4}$$

for all $F \in L$ and all $W \in K$. Moreover, (1.3) and the definitions (1.1) and (1.2) of the equivalence classes W and F also imply that

$$W_1 = W_2 \text{ iff } \mu(F,W_1) = \mu(F,W_2) \text{ for all } F \in L \,, \tag{1.5}$$

and

$$F_1 = F_2 \text{ iff } \mu(F_1,W) = \mu(F_2,W) \text{ for all } W \in K \,. \tag{1.6}$$

This means, in particular, that ensembles in different states $W_1 \neq W_2$, as well as apparatuses measuring different effects $F_1 \neq F_2$, lead to a different "statistics", and can thus be distinguished experimentally.

One of the main goals of Ludwig's approach is the derivation of the mathematical structure of quantum mechanics from suitable and physically meaningful postulates for the sets K and L and the probability function μ. (The classical example for such an "axiomatic" approach is thermodynamics, as based on the first and second law.) Again there is no room here to discuss this in any detail. (See [2].) Rather than deriving the usual Hilbert space formalism of quantum mechanics, we shall thus take for granted here that the basic quantities and relations of the theory may be represented mathematically in terms of operators on a (complex, separable) Hilbert space H, which we call the state space of the system considered. By restricting, moreover, our attention to sufficiently "simple" systems (like, e.g., single electrons), we shall also avoid possible complications which otherwise could arise from superselection rules.

Accordingly, we shall assume - as usual - that

(i) states may be represented by density operators W on H,

(ii) projection operators E on H represent effects, and

(iii) the probability for the occurrence of such an effect E in a state W is given by

$$\mu(E,W) = \mathrm{tr}(EW) \;, \tag{1.7}$$

with tr denoting the operator trace.

For simplicity of notation, we have avoided the introduction of different symbols for states and effects on the one hand, and the operators on H representing them on the other hand. Thus, in (i), (ii), and on the right hand side of (1.7), W and E denote operators, whereas in (iii) and on the left hand side of (1.7) the same letters stand for the corresponding physical quantities (i.e., equivalence classes of instruments).

Statement (i) shall mean, more precisely, that to every state there corresponds a unique density operator, and vice versa. With this assumption, the set K of states may be identified with the set K(H) of density operators on H, i.e., of non-negative (and thus) Hermitean operators W with unit trace:

$$W = W^* \geq 0 \;, \quad \mathrm{tr}W = 1 \;. \tag{1.8}$$

K(H) is a subset of the trace class $B(H)_1$, consisting of all operators T on H for which the trace norm

$$\|T\|_1 = \mathrm{tr}[(T^*T)^{1/2}] \tag{1.9}$$

is finite. For the properties of $B(H)_1$ – to be used later on – see, e.g., [3], Ch. 1, or [4]. As a subset of $B(H)_1$ – or alternatively, of its Hermitean part $B(H)_1^h$, consisting of all Hermitean trace class operators – K(H) is convex; i.e., with $0 \leq \lambda \leq 1$ and $W_1, W_2 \in K(H)$,

$$W = \lambda W_1 + (1-\lambda)W_2 \tag{1.10}$$

also belongs to K(H). Physically, this expresses the possibility of state mixing. Assume an ensemble of $N \gg 1$ systems to be prepared by using $N_1 = \lambda N$ times a preparing instrument w_1 and $N_2 = (1-\lambda)N$ times another one, w_2. A prescription of this type, with λ, w_1 and w_2 fixed, defines a new preparation procedure w, called the mixing of the prestates w_1 and w_2 with relative weights λ and $1 - \lambda$. If an effect apparatus f is applied to the ensemble, it will be triggered $\tilde{\mu}(f,w_1)N_1$ times by the subensemble prepared by the instrument w_1, and $\tilde{\mu}(f,w_2)N_2$ times by the N_2 systems prepared by the instrument w_2. Thus the probability for the triggering of f in the mixed prestate w is

$$\tilde{\mu}(f,w) = (\tilde{\mu}(f,w_1)N_1 + \tilde{\mu}(f,w_2)N_2)/N$$

$$= \lambda\tilde{\mu}(f,w_1) + (1-\lambda)\tilde{\mu}(f,w_2) \ .$$

This relation implies that replacing w_1 and w_2 by equivalent w_1' and w_2' leads to a w' equivalent to w, and that an analogous relation also holds true for equivalence classes:

$$\mu(F,W) = \lambda\mu(F,W_1) + (1-\lambda)\mu(F,W_2) \ . \tag{1.11}$$

If applied, in particular, to effects described by projection operators E, the last relation, together with (1.7), leads to

$$\mathrm{tr}(EW) = \lambda\mathrm{tr}(EW_1) + (1-\lambda)\mathrm{tr}(EW_2)$$

$$= \mathrm{tr}[E(\lambda W_1 + (1-\lambda)W_2)] \ . \tag{1.12}$$

As E is arbitrary, this implies (1.10). (Take, e.g., $E = |f\rangle\langle f|$, the projection operator onto the one-dimensional subspace spanned by an arbitrary unit vector $f \in H$. Then (1.12) becomes $(f,Af) = 0$, with $A = W - \lambda W_1 - (1-\lambda)W_2$. Since A is linear, we also have $(f,Af) = 0$ for vectors f of arbitrary length, and the polarization identity

$$4(f,Ag) = ((f+g),A(f+g)) - ((f-g),A(f-g))$$

$$+ i((f-ig),A(f-ig)) - i((f+ig),A(f+ig)) \tag{1.13}$$

yields $(f,Ag) = 0$ for all $f, g \in H$, i.e., $A = 0$.)

A state W satisfying (1.10) is thus called, in view of this physical interpretation, a mixture of the states W_1 and W_2. States W which are not proper mixtures, i.e., which can not be represented in the form (1.10) with $W_1 \neq W_2$ and $0 < \lambda < 1$, are called pure states. A pure state, as is well-known, corresponds to a one-dimensional projection operator,

$$W = | f \rangle\langle f | , \quad \| f \| = 1 , \tag{1.14}$$

and is usually represented by the unit ray $\{e^{i\alpha} f\}$ spanned by the "state vector" f. In this case, Eq. (1.7) reads

$$\mu(E,W) = (f,Ef) .$$

Although sufficient for many purposes, it is nevertheless not very satisfactory to restrict the discussion of quantum states to pure states only. For even by disregarding as "artificial" preparation procedures like the above-described state mixing, one would not get rid of state mixtures, since there are also "simple" preparation procedures, with single instruments, which nevertheless do not produce pure states.

Statement (ii) above is meant here to imply that every projection operator E on H describes an effect. Usually one also assumes that, vice versa, every yes-no measurement can be described by a projection operator E on H. As we shall see, however, there are good reasons to modify this assumption by admitting a larger set L(H) of operators F on H as describing effects. This set L(H) consists of all operators F which are non-negative and (therefore) Hermitean, and bounded from above by the unit operator:

$$0 \leq F = F^* \leq 1 . \tag{1.15}$$

The set of projection operators E is a proper subset of L(H). Statement (iii) is then generalized to arbitrary effects F by requiring

$$\mu(F,W) = tr(FW) \tag{1.16}$$

for all effects $F \in L \equiv L(H)$ and all states $W \in K = K(H)$.

The set L(H) is also convex, i.e., containing with F_1 and F_2 also

$$F = \lambda F_1 + (1-\lambda)F_2 \tag{1.17}$$

for all real λ between 0 and 1. A particular effect apparatus f corresponding to such an effect F could be obtained, as in the analogous case of Eq. (1.10), by "mixing" two effect apparatuses f_1 and f_2 corresponding to the effects F_1 and F_2, respectively: Applying the "mixed" apparatus f to an ensemble of N microsystems means that λN times the apparatus f_1, and $(1-\lambda)N$ times the apparatus f_2 has to be applied. Since the set of projection operators is not convex, such "mixing" of effects would not be allowed if every effect, as in the usual formulation of quantum mechanics, were to be described by a projection operator. Now, indeed, the above prescription for measuring the effect (1.17) does not look very natural, so that one might still hope that at least all sufficiently "simple" effect apparatuses correspond to projection operators.

As we shall try to show later on, however, this hope is not justified. On the contrary, projection operators describe only very particular effects - called "decision effects" by Ludwig [2] - which moreover are rather unlikely to be realized at all in actual experiments. In this respect, the convex set L(H) of effects is quite similar to the convex set K(H) of states, which also contains, as a non-convex subset, the set of pure states. But whereas in the latter case there is general agreement that the pure states form only a subset of K(H), and that many actual preparing instruments - or perhaps even most of them - yield proper mixtures rather than pure states, corresponding statements about the particular rôle of the decision effects (projection operators) in L(H) are much less popular.

Pure states can be characterized operationally by the fact that they cannot be prepared as proper state mixtures (see above). One might then ask whether and how, similarly, the decision effects E could be distinguished physically from more general effects F in L(H). According to Ludwig [2], decision effects are indeed distinguished as the "most sensitive" effects in suitable subsets of L(H). We shall not discuss this here, but will discover another characteristic property of decision effects in §6.

§2 Operations

We shall now introduce, as another fundamental notion, the new concept of an "operation". Like "state" and "effect", this concept shall also be defined operationally. It was first used in quantum theory by Haag and Kastler [5].

Assume an ensemble of $N >> 1$ microsystems has been prepared by a preparing instrument w, so that the ensemble is in the corresponding state W. Assume, moreover, that an effect apparatus f is applied to each microsystem, and that each microsystem is still present - and thus available for further experiments - after its interaction with the apparatus f.

(Actually there are many instruments f, which do not satisfy this assumption, but instead "absorb" or "destroy" the microsystem. To such instruments the subsequent discussion does not apply. On the other hand, there are many effect apparatuses f as well which act "non-destructively" in the above sense. As a given effect F can be measured by many different instruments f in the corresponding equivalence class, it appears even reasonable to expect that every effect F may be measured "non-destructively" by a suitable apparatus f.)

In the situation described, it is allowed to consider the N microsystems after their interaction with the apparatus f as a new ensemble, and to ascribe to this ensemble a quantum state $\tilde{W}$, which in general will be different from the original state W as prepared by the instrument w. This just amounts to considering the combination of the original preparing instrument w and the given effect apparatus f as a single new preparing instrument $\tilde{w}$, which then belongs to a certain equivalence class of preparing instruments defining the new state $\tilde{W}$. This state, clearly, depends on both the preparing instrument w and the effect apparatus f constituting the new preparing instrument $\tilde{w}$.

We will show, however, that combining the given effect apparatus f with another preparing instrument w' equivalent to w leads to a new preparing instrument $\tilde{w}'$ equivalent to $\tilde{w}$ - or, in other words: that the new state $\tilde{W}$ depends on f and the equivalence class of w (i.e., the initial state W) only. To show this, assume $\tilde{w}$ and $\tilde{w}'$ to be inequivalent. Accordingly, there exists at least one effect apparatus g, such that

$$\tilde{\mu}(g,\tilde{w}) \neq \tilde{\mu}(g,\tilde{w}') .$$

It is perfectly legitimate, however, to consider the combination of the instruments f and g as another apparatus h, which is defined to give the result "yes" ("no") if its "g part" is triggered (not triggered), irrespective of the response of its "f part". If reformulated in terms of this effect apparatus h, the above inequality becomes

$$\tilde{\mu}(h,w) \neq \tilde{\mu}(h,w') ,$$

which is a contradiction since w and w' are equivalent.

Whereas thus the new state $\tilde{W}$ depends on the initial preparing instrument w only through its equivalence class, the analogous statement for the effect apparatus f does not hold. Indeed, different apparatuses f within the same equivalence class commonly yield different states $\tilde{W}$, as will be shown later.

An apparatus f applied in the way just described, thereby transforming a given ensemble of N systems in some initial state W into another ensemble - again consisting of N systems - in a new state $\tilde{W}$ depending on f and W, is said to perform a non-selective operation. Keeping the apparatus f fixed and varying the initial state W to which it is applied, a non-selective operation thus generates a certain mapping

$$\tilde{\phi} : W \to \tilde{W} = \tilde{\phi} W \tag{2.1}$$

of the set K(H) of normalized density operators (states) into itself. The mapping $\tilde{\phi}$ thus describes all possible state changes induced by the given effect apparatus f. As the mathematical counterpart of the physical procedure described above, this mapping $\tilde{\phi}$ will also be called here a "non-selective operation".

By using the same effect apparatus f in a slightly different way, one can also perform a selective operation, defined operationally as follows. When applied to an ensemble of N >> 1 systems in a state W, the apparatus f will be triggered in $N_+ = \tilde{\mu}(f,w)N$ cases. By selecting, after their interaction with the apparatus f, the N_+ microsystems which have triggered f, while disregarding the remaining $N_- = N - N_+$ systems, we arrive at another ensemble, now consisting of N_+ microsystems. (With N, N_+ can also be made arbitrarily large unless $\tilde{\mu}(f,w) = 0$, an exceptional case to be considered separately.) This ensemble is also in a well-defined new state $\hat{W}$, depending again on both applied instruments, w and f, since as above the combination of the instruments w and f - now applied in a somewhat different way, however - may be considered as a new preparing instrument $\hat{w}$ belonging to a well-defined equivalence class $\hat{W}$.

As before, the new state $\hat{W}$ depends on the equivalence class W of w only, rather than on the particular w chosen from it. To show this, consider the application of an arbitrary effect apparatus g to the N_+ systems prepared by the procedure $\hat{w}$. Denoting by N_{++} the number of cases in which the apparatus g is triggered by this ensemble, we have by definition

$$\tilde{\mu}(g,\hat{w}) = N_{++}/N_+ \ . \tag{2.2}$$

Consider again the combination of the effect apparatuses f and g as a new effect apparatus $\hat{h}$, but define now the effect to be measured by $\hat{h}$

to occur if and only if both the f and g part of the combined apparatus are triggered. Since thus used in a different way, the combination of f and g now becomes an effect apparatus $\hat{h}$ which is different from the apparatus h considered above. (For a more detailed discussion of combined yes-no measurements see §6.) In a single experiment of the kind considered, the successive triggering of both the f and g apparatus may be interpreted either as a triggering of g by a system prepared by the procedure $\hat{w}$, or else as a triggering of $\hat{h}$ by a system prepared by the instrument w. The first interpretation corresponds to Eq. (2.2), whereas the second one yields the relation

$$N_{++} = \tilde{\mu}(\hat{h},w)N \ .$$

Since

$$N_{+} = \tilde{\mu}(f,w)N \ ,$$

we may rewrite (2.2) in the form

$$\tilde{\mu}(g,\hat{w}) = \tilde{\mu}(\hat{h},w)/\tilde{\mu}(f,w) \ . \tag{2.3}$$

The right hand side of this is unchanged if w is replaced by an arbitrary w' in the same equivalence class W. Therefore $\tilde{\mu}(g,\hat{w})$ is also unchanged, for arbitrary g; i.e., this replacement leads to the same state $\hat{W}$.

If $\tilde{\mu}(f,w) = 0$, we have $N_{+} = \tilde{\mu}(f,w)N = 0$ as well; i.e., the combined use of w and f according to the selection prescription described above does not really lead to a preparation procedure $\hat{w}$. Accordingly, a "new state" $\hat{W}$ resulting from the "selective operation" can not - and need not - be defined in this case.

In contrast to a non-selective operation, a selective operation, applied to N microsystems in a state W, does not always lead to N but in general only to $N_{+} < N$ systems in the new state $\hat{W}$. The fraction N_{+}/N is called - quite legitimately, of course - the transition probability of the state change $W \to \hat{W}$. (If $N_+ = 0$, no final state $\hat{W}$ exists, and the "transition probability" is zero.) As this transition probability coincides, by definition of the selection procedure, with the probability $\tilde{\mu}(f,w) = \mu(F,W)$ of the effect F in the initial state W, it depends on the effect apparatus f up to equivalence only. As already stated above for non-selective operations, however, a corresponding statement for the final state $\hat{W}$ itself will turn out to be wrong.

Mathematically, a selective operation could be identified with the mapping $W \to \hat{W}$ of initial into final states induced by it, as we did before for a non-selective operation. This would have some disadvantages, however. First, such a mapping would not be defined on all of K(H), since $\hat{W}$ does not exist if $\tilde{\mu}(f,w) = \mu(F,W) = 0$. Second, this mapping would not specify the transition probabilities, which therefore would have to be given separately. We therefore associate with a selective operation a slightly different mapping ϕ – also called a "selective operation" in the following – which is defined, for arbitrary initial states $W \in K(H)$, by

$$\phi W = \begin{cases} \mu(F,W)\hat{W} & \text{if } \mu(F,W) \neq 0 \\ 0 & \text{if } \mu(F,W) = 0 \end{cases} \tag{2.4}$$

in terms of the final states $\hat{W}$ and the transition probabilities $\mu(F,W)$. Since $\mathrm{tr}\hat{W} = 1$ by definition, both the transition probability

$$\mu(F,W) = \mathrm{tr}(\phi W) \tag{2.5}$$

and – if it exists – the final state

$$\hat{W} = \phi W/\mathrm{tr}(\phi W) \tag{2.6}$$

can then be calculated for an arbitrary initial state W from the mapping ϕ.

According to its definition (2.4), ϕ does not map the set of states K(H) into itself, but rather into the set $B(H)_1^+$ of non-negative (and thus Hermitean) trace class operators (or "unnormalized" density operators) with $tr(\phi W) \leq 1$. In view of the additional physical information provided by Eq. (2.5), however, this is an advantage rather than a disadvantage. The most important motivation for the definition (2.4) is, however, that ϕ is convex-linear, i.e., satisfying

$$\phi(\lambda W_1 + (1-\lambda)W_2) = \lambda\phi W_1 + (1-\lambda)\phi W_2 \tag{2.7}$$

for arbitrary W_1, $W_2 \in K(H)$ and all real λ with $0 \leq \lambda \leq 1$.

For the proof of (2.7), assume first that the transition probabilities $tr(\phi W_1)$ and $tr(\phi W_2)$ are both non-vanishing. As already discussed in §1, an ensemble of N systems in the mixed state

$$W = \lambda W_1 + (1-\lambda)W_2$$

may be prepared by applying $\lambda N = N_1$ times an instrument w_1 preparing the state W_1, and $(1-\lambda)N = N_2$ times another instrument w_2 preparing W_2. (Since (2.7) is trivially satisfied for $\lambda = 0$ or 1, we also assume $0 < \lambda < 1$.) Applying the selective operation ϕ in this particular case then amounts to replacing the preparing instruments w_1 and w_2 by $\hat{w}_1$ and $\hat{w}_2$, respectively, by adding to each of them the effect apparatus f and the above selection prescription. The state $\hat{W} = \phi W/tr(\phi W)$ of the ensemble after the operation ϕ is thus some mixture,

$$\hat{W} = \hat{\lambda}\hat{W}_1 + (1-\hat{\lambda})\hat{W}_2 ,$$

of the states $\hat{W}_1 = \phi W_1/tr(\phi W_1)$ and $\hat{W}_2 = \phi W_2/tr(\phi W_2)$ prepared by $\hat{w}_1$ and $\hat{w}_2$, respectively.

The instrument w_1 prepares N_1 systems, $N_{1+} = tr(\phi W_1)N_1$ of them surviving the selection procedure performed with the apparatus f, and thus leaving the preparing instrument $\hat{\mathcal{W}}_1$. Similarly, $\hat{\mathcal{W}}_2$ prepares $N_{2+} = \mathrm{tr}(\phi W_2)N_2$ systems. The final ensemble thus contains $N_+ = N_{1+} + N_{2+}$ systems; so the transition probabilty of the state change $W \to \hat{W}$ is

$$\mathrm{tr}(\phi W) = N_+/N .$$

Moreover, the weight factors in the mixed state $\hat{W}$ become

$$\hat{\lambda} = \frac{N_{1+}}{N_+} = \frac{\mathrm{tr}(\phi W_1)N_1}{\mathrm{tr}(\phi W)N} = \frac{\mathrm{tr}(\phi W_1)}{\mathrm{tr}(\phi W)}\,\lambda$$

and

$$1 - \hat{\lambda} = \frac{\mathrm{tr}(\phi W_2)}{\mathrm{tr}(\phi W)}\,(1 - \lambda) .$$

This, finally, implies

$$\begin{aligned}\phi W &= \mathrm{tr}(\phi W)\hat{W} = \mathrm{tr}(\phi W)(\hat{\lambda}\hat{W}_1 + (1-\hat{\lambda})\hat{W}_2) \\ &= \lambda \mathrm{tr}(\phi W_1)\hat{W}_1 + (1-\lambda)\mathrm{tr}(\phi W_2)\hat{W}_2 \\ &= \lambda\phi W_1 + (1-\lambda)\phi W_2 ,\end{aligned}$$

i.e., Eq. (2.7).

The same result is also obtained in the exceptional cases

$$\mathrm{tr}(\phi W_1) \neq 0 , \quad \mathrm{tr}(\phi W_2) = 0$$

and

$$\mathrm{tr}(\phi W_1) = \mathrm{tr}(\phi W_2) = 0 .$$

As $\phi W_2 \geq 0$ by definition, $\mathrm{tr}(\phi W_2) = 0$ implies $\phi W_2 = 0$. In the first case we therefore get $N_{2+} = 0$, and thus $\hat{\lambda} = 1$ and $\mathrm{tr}(\phi W) = N_{1+}/N = (N_1/N)(N_{1+}/N_1) = \lambda \mathrm{tr}(\phi W_1)$, which implies (2.7):

$$\phi W = \mathrm{tr}(\phi W)\hat{W}_1 = \lambda\phi W_1 = \lambda\phi W_1 + (1-\lambda)\phi W_2 .$$

In the second case we have $\phi W_1 = \phi W_2 = 0$ and - since $N_+ = 0$, and thus $tr(\phi W) = 0$ - also $\phi W = 0$, so that (2.7) holds trivially.

Before continuing with our investigation of the mathematical properties of the mapping ϕ, we shall illustrate the general discussion by a specific example. In textbooks of quantum mechanics one usually finds some version of the "wave packet reduction formula", or "projection postulate", which in the language used here can be formulated as follows: If a decision effect E is measured in an ensemble of N systems in state W, then the $N_+ = tr(EW)N$ systems which have triggered the effect E are, after the measurement, in the new state $\hat{W} = EWE/tr(EW)$.

Since in the case considered the tansition probability is tr(EW), (2.4) leads to

$$\phi W = EWE \; . \tag{2.8}$$

The corresponding non-selective operation is

$$\tilde{\phi} W = EWE + (1-E)W(1-E) \; . \tag{2.9}$$

Namely, the projection postulate applied to the effect 1-E, describing the non-occurrence of E, implies that, after the measurement, the $N_- = tr((1-E)W)N$ systems which have not triggered the effect E are in the state $\hat{W}' = (1-E)W(1-E)/tr((1-E)W)$. Therefore, if no selection is made, the state $\tilde{W} = \tilde{\phi} W$ after the E measurement is a mixture of $\hat{W}$ and $\hat{W}'$ with the weight factors tr(EW) and tr((1-E)W), respectively, and is thus given by (2.9).

It is commonly admitted that, in practice, there are also E measurements which do not lead to the final states $\hat{W}$ or $\tilde{W}$ given by (2.8) and (2.9). For this reason, a measurement satisfying the projection postulate is sometimes called an "ideal measurement". We shall indeed show that such "ideal measurements" are only a very particular type of operations, and that actual measurements of decision effects E are not very likely to be of this type. Besides this, to effects F which are not decision effects Eqs. (2.8) and (2.9) would not be applicable anyway. Nevertheless, the examples (2.8) and (2.9) should be kept in mind as simple examples illustrating the general discussion, to which we now return.

By virtue of (2.7), the mapping $\phi : K(H) \to B(H)_1^+$ can be extended in a unique way to a mapping - for simplicity also denoted by ϕ - of the trace class $B(H)_1$ into itself, which is complex-linear and positive. I.e., the extended mapping $\phi : B(H)_1 \to B(H)_1$ satisfies

$$\phi(aT + bS) = a\phi T + b\phi S \tag{2.10}$$

for all $T, S \in B(H)_1$ and all complex numbers a and b; and

$$\phi T \geq 0 \quad \text{if} \quad T \geq 0 \ . \tag{2.11}$$

By (2.10) and (2.11) - or else by its explicit construction, cf. Eq. (2.18) below - ϕ is also real, i.e.,

$$\phi T^* = (\phi T)^* \ . \tag{2.12}$$

To see this, note that any Hermitean T is of the form $T_+ - T_-$ with $T_\pm \geq 0$. By (2.11), the operators $\phi T_\pm$ are ≥ 0, and thus Hermitean, and by (2.10), then, $\phi T = \phi T_+ - \phi T_-$ is also Hermitean. A non-Hermitean T may be written as $T_1 + iT_2$ with Hermitean T_1 and T_2, so that, again by (2.10),

$$\phi T^* = \phi(T_1 - iT_2) = \phi T_1 - i\phi T_2 = (\phi T_1 + i\phi T_2)^* = (\phi T)^* \ .$$

Although the procedure of extending ϕ from K(H) to $B(H)_1$ is quite standard, we shall sketch it here for the reader's convenience. It is done in three steps.

In a first step, the original mapping ϕ is extended to a mapping ϕ_+ of the cone $B(H)_1^+$ of "unnormalized" density operators into itself, which is also convex-linear (or rather "positive-linear"); i.e.,

$$\phi_+(T + T') = \phi_+T + \phi_+T' \tag{2.13}$$

and

$$\phi_+aT = a\phi_+T \tag{2.14}$$

for all $T, T' \in B(H)_1^+$ and all numbers $a \geq 0$. To achieve this, set $\phi_+T = 0$ if $T = 0$, and

$$\phi_+T = trT\cdot\phi(T/trT) \tag{2.15}$$

if $T \neq 0$, the last expression being well-defined since $T/trT \in K(H)$ if $0 \neq T \in B(H)_1^+$. From this, (2.14) follows trivially. Moreover, with $T, T' \in B(H)_1^+$ (and neither of them equal to zero, since otherwise (2.13) holds trivially), we get

$$V \underset{df.}{=} \frac{T + T'}{tr(T+T')} = \frac{trT}{tr(T+T')}\frac{T}{trT} + \frac{trT'}{tr(T+T')}\frac{T'}{trT'}$$

$$= \lambda W + (1-\lambda)W'$$

with V, $W = T/trT$ and $W' = T'/trT' \in K(H)$ and $0 < \lambda < 1$, so that, by (2.7) and (2.15),

$$\phi_+(T + T') = tr(T+T')\cdot\phi V = tr(T+T')(\lambda\phi W + (1-\lambda)\phi W')$$

$$= trT\cdot\phi W + trT'\cdot\phi W' = \phi_+T + \phi_+T' ,$$

which proves (2.13).

In a second step, the mapping ϕ_+ is extended to a real-linear mapping of the real space $B(H)_1^h$ of Hermitean trace class operators into itself. If $T = T^* \in B(H)_1^h$, we have $T = T_+ - T_-$ with $T_\pm \in B(H)_1^+$; take, e.g., $T_{\pm} = (|T| \pm T)/2$ with $|T| = (T^2)^{1/2}$. Such decompositions are not unique, however; e.g., with arbitrary $c > 0$ we also have $T = (1+c)T_+ - (T_- + cT_+)$. Nevertheless, ϕ_r defined by

$$\phi_r T = \phi_+ T_+ - \phi_+ T_- \tag{2.16}$$

with any of these decompositions of T is unique. For, if $T = T_+ - T_- = S_+ - S_-$ with $T_{\pm}, S_{\pm} \in B(H)_1^+$, we have $T_+ + S_- = S_+ + T_- \in B(H)_1^+$ and thus, by (2.13),

$$\phi_+(T_+ + S_-) = \phi_+ T_+ + \phi_+ S_- = \phi_+(S_+ + T_-) = \phi_+ S_+ + \phi_+ T_- ,$$

or

$$\phi_+ T_+ - \phi_+ T_- = \phi_+ S_+ - \phi_+ S_- .$$

There remains to prove real-linearity of ϕ_r, i.e.,

$$\phi_r(aT) = a\phi_r T , \quad \phi_r(T + T') = \phi_r T + \phi_r T' \tag{2.17}$$

for all $T, T' \in B(H)_1^h$ and all real a. We have $aT = S_+ - S_-$ with

$$S_+ = \pm aT_{\pm} , \quad S_- = \pm aT_{\mp} \in B(H)_1^+ ,$$

the upper (lower) signs being valid if $a \geqslant 0$ $(a < 0)$, and thus, with (2.14) and (2.16),

$$\phi_r(aT) = \phi_+ S_+ - \phi_+ S_- = \pm a\phi_+ T_{\pm} \mp a\phi_+ T_{\mp}$$

$$= a(\phi_+ T_+ - \phi_+ T_-) = a\phi_r T .$$

Moreover, with the decomposition

$$T + T' = (T_+ + T'_+) - (T_- + T'_-)$$

we obtain from (2.13) and (2.16)

$$\phi_r(T + T') = \phi_+(T_+ + T'_+) - \phi_+(T_- + T'_-)$$

$$= \phi_+T_+ - \phi_+T_- + \phi_+T'_+ - \phi_+T'_- = \phi_rT + \phi_rT' .$$

Finally, ϕ_r is extended to a mapping ϕ_c of $B(H)_1$ into itself by defining, with the unique decomposition $T = T_1 + iT_2$ of $T \varepsilon B(H)_1$ into Hermitean T_1 and T_2,

$$\phi_cT = \phi_rT_1 + i\phi_rT_2 . \tag{2.18}$$

With $T' = T'_1 + iT'_2$, and using (2.17), we then get

$$\phi_c(T + T') = \phi_c(T_1 + T'_1 + i(T_2 + T'_2))$$

$$= \phi_r(T_1 + T'_1) + i\phi_r(T_2 + T'_2)$$

$$= \phi_rT_1 + i\phi_rT_2 + \phi_rT'_1 + i\phi_rT'_2$$

$$= \phi_cT + \phi_cT' . \tag{2.19}$$

Moreover, with real α and β,

$$(\alpha + i\beta)T = (\alpha T_1 - \beta T_2) + i(\alpha T_2 + \beta T_1) ,$$

and thus, again by (2.17),

$$\phi_c((\alpha + i\beta)T) = \phi_r(\alpha T_1 - \beta T_2) + i\phi_r(\alpha T_2 + \beta T_1)$$

$$= \alpha\phi_rT_1 - \beta\phi_rT_2 + i\alpha\phi_rT_2 + i\beta\phi_rT_1$$

$$= (\alpha + i\beta)(\phi_rT_1 + i\phi_rT_2)$$

$$= (\alpha + i\beta)\phi_cT . \tag{2.20}$$

Eqs. (2.19) and (2.20) are equivalent to (2.10), the complex-linearity of ϕ_c.

If $T = T^*$, (2.18) yields $\phi_c T \equiv \phi_r T$. Likewise, (2.16) implies $\phi_r T = \phi_+ T$ if $T \in B(H)_1^+$, and (2.15) implies $\phi_+ T = \phi T$ if $T \in K(H)$. Thus ϕ_+, ϕ_r and ϕ_c are really extensions of the original mapping $\phi : K(H) \to B(H)_1^+$.

Positivity of ϕ_c (Eq. (2.11)) follows immediately since ϕ_c reduces to ϕ_+ on $B(H)_1^+$, and $\phi_+ T \geq 0$ by (2.15). As announced before (and already done in (2.10) and (2.11)), we shall omit the suffix in ϕ_c, thus identifying an operation with the extended mapping $\phi : B(H)_1 \to B(H)_1$, from now on.

We shall now prove that the mapping ϕ is continuous with respect to the trace norm:

$$\|\phi T\|_1 \leq C \, \|T\|_1 \, , \quad \text{with} \quad C = \sup_{W \in K(H)} \operatorname{tr}(\phi W) \leq 1 \, . \tag{2.21}$$

We recall here the definition (1.9),

$$\|T\|_1 = \operatorname{tr}|T| \, , \qquad |T| = (T^*T)^{1/2} ,$$

and some properties of the trace norm $\|\cdot\|_1$. (Compare [3], Ch. 1, and [4].) Denoting by $B(H)$ the algebra of all bounded operators on H and by $\|\cdot\|$ the operator norm, we have

$$|\operatorname{tr}(XT)| \leq \|X\| \, \|T\|_1$$

for arbitrary $X \in B(H)$, $T \in B(H)_1$, and

$$\|T\|_1 = \sup_{\|X\|=1} |\operatorname{tr}(XT)| ,$$

$$\|X\| = \sup_{\|T\|_1=1} |\operatorname{tr}(XT)| = \sup_{W \in K(H)} |\operatorname{tr}(XW)| .$$

The existence of

$$C = \sup_{W \in K(H)} \operatorname{tr}(\phi W)$$

as required in (2.21), as well as the statement $C \leq 1$, follow immediately from the physical interpretation of $\operatorname{tr}(\phi W)$ as transition probability. (That $\operatorname{tr}(\phi W)$ is bounded from above, however, is already guaranteed by the positivity of ϕ. See, e.g., [3], Ch. 2.)

In order to prove (2.21), we start with the decomposition

$$T = T_+ - T_- = \operatorname{tr}T_+ \cdot W_+ - \operatorname{tr}T_- \cdot W_- , \tag{2.22}$$

with

$$T_\pm = (|T| \pm T)/2 , \quad W_\pm = T_\pm / \operatorname{tr}T_\pm ,$$

of an arbitrary Hermitean $T \in B(H)_1$. (If, e.g., $T_+ = 0$, take W_+ arbitrary.) Then

$$|T| = T_+ + T_- , \quad \|T\|_1 = \operatorname{tr}T_+ + \operatorname{tr}T_- , \tag{2.23}$$

and

$$\phi T = \operatorname{tr}T_+ \cdot \phi W_+ - \operatorname{tr}T_- \cdot \phi W_- .$$

Therefore we get, for all $T \in B(H)_1^h$,

$$\begin{aligned} \|\phi T\|_1 &= \sup_{\|X\|=1} |\operatorname{tr}(X \cdot \phi T)| \\ &\leq \sup_{\|X\|=1} (\operatorname{tr}T_+ |\operatorname{tr}(X \cdot \phi W_+)| + \operatorname{tr}T_- |\operatorname{tr}(X \cdot \phi W_-)|) \\ &\leq C(\operatorname{tr}T_+ + \operatorname{tr}T_-) = C \|T\|_1 , \end{aligned} \tag{2.24}$$

since

$$\operatorname{tr}(X \cdot \phi W_\pm) \leq \|X\| \, \|\phi W_\pm\|_1 = \operatorname{tr}(\phi W_\pm) \leq C .$$

Denote by $B(H)^h$ the set of all bounded Hermitean operators on H, and consider a fixed $X \in B(H)^h$. Then

$$\hat{x}(T) \underset{df.}{\equiv} tr(X \cdot \phi T) ,$$

with $T \in B(H)_1^h$ arbitrary, defines a real and real-linear functional on $B(H)_1^h$. Moreover, by (2.24),

$$|\hat{x}(T)| = |tr(X \cdot \phi T)| \leq \|X\| \; \|\phi T\|_1 \leq C \; \|X\| \; \|T\|_1 ;$$

i.e., $\hat{x}(T)$ is continuous with respect to the trace norm. Therefore it is of the form

$$\hat{x}(T) = tr(\hat{X}T)$$

with a unique operator

$$\hat{X} \underset{df.}{\equiv} \phi^* X \in B(H)^h .$$

Here we have made use of the fact that the real Banach space $B(H)^h$ (with the norm $\|\cdot\|$) is the dual of the real Banach space $B(H)_1^h$ (with the norm $\|\cdot\|_1$), in the following sense: Every (real-)linear (real) functional $x(T)$ on $B(H)_1^h$ which is continuous with respect to the trace norm is of the form $tr(XT)$ with a unique $X \in B(H)^h$; vice versa, $tr(XT)$ with an arbitrary $X \in B(H)^h$ defines such a functional $x(T)$ on $B(H)_1^h$, and the norm

$$\|x\| = \sup_{\|T\|_1 = 1} |x(T)|$$

of this functional coincides with $\|X\|$ (cf. [3], Ch. 1).

Varying X in the above construction, we obtain a mapping $\phi^* : X \to \hat{X}$ of $B(H)^h$ into itself. By definition,

$$tr(X \cdot \phi T) = tr(\phi^* X \cdot T) \tag{2.25}$$

for all $X \in B(H)^h$, $T \in B(H)_1^h$. In this sense, ϕ^* is the adjoint of the mapping ϕ of $B(H)_1^h$ into itself. (Strictly speaking, we are dealing here with the restriction ϕ_r of ϕ to $B(H)_1^h$; thus ϕ^* would better be called ϕ_r^*). By (2.25), the mapping ϕ^* is real-linear, and can thus be extended in a unique way to a complex-linear mapping of B(H) into itself, which for simplicity is also denoted by ϕ^*: For any $X = X_1 + iX_2 \in B(H)$, with (unique) $X_{1,2} \in B(H)^h$, define

$$\phi^* X = \phi^* X_1 + i\phi^* X_2 \ .$$

Complex-linearity then follows as above for the analogous extension ϕ_c of ϕ_r.

With this extension of ϕ^*, then, (2.25) becomes valid for arbitrary $X \in B(H)$ and $T \in B(H)_1$ - or, in other words: the complex-linear mapping ϕ^* of the complex Banach space B(H) is the adjoint of the complex-linear mapping ϕ ($\equiv \phi_c$) of the complex Banach space $B(H)_1$. (The complex Banach space B(H) is the dual of the complex Banach space $B(H)_1$ in the same sense as explained above for the real Banach spaces $B(H)_1^h$ and $B(H)^h$, cf. [3], Ch. 1, or [4].) The extended validity of (2.25) follows easily by inserting arbitrary $X = X_1 + iX_2 \in B(H)$ and $T = T_1 + iT_2 \in B(H)_1$ into $tr(X \cdot \phi T)$ and $tr(\phi^* X \cdot T)$, and comparing the resulting expressions; they are equal since, as is already known, $tr(X_i \cdot \phi T_j) = tr(\phi^* X_i \cdot T_j)$ for $i,j = 1,2$.

The mapping ϕ^* is continuous with respect to the operator norm:

$$\|\phi^* X\| = \sup_{W \in K(H)} |tr(\phi^* X \cdot W)| = \sup_{W} |tr(X \cdot \phi W)|$$

$$\leq \sup_{W} \|X\| \ \|\phi W\|_1 = \|X\| \sup_{W} tr(\phi W) = C \ \|X\| \ . \qquad (2.26)$$

Thus, finally,

$$\|\phi T\|_1 = \sup_{\|X\|=1} |tr(X \cdot \phi T)| = \sup_{\|X\|=1} |tr(\phi^* X \cdot T)|$$

$$\leq \sup_{\|X\|=1} \|\phi^* X\| \ \|T\|_1 \leq C \ \|T\|_1 \ ,$$

which completes the proof of (2.21).

Like ϕ, its adjoint ϕ^* is positive:

$$\phi^* X \geq 0 \quad \text{if} \quad X \geq 0 \ , \tag{2.27}$$

and - therefore, or by the explicit construction above - real:

$$\phi^* X^* = (\phi^* X)^* \ .$$

For if $X \geq 0$, we have

$$\mathrm{tr}(\phi^* X \cdot W) = \mathrm{tr}(X \cdot \phi W) \geq 0$$

for all $W \in K(H)$, which implies $\phi^* X \geq 0$. (Take, e.g., $W = | f >< f |$ with an arbitrary unit vector $f \in H$.)

The mapping ϕ^* also provides an explicit representation of the effect F belonging to the operation ϕ. By (2.5) and (2.25), we have

$$\mu(F,W) = \mathrm{tr}(\phi W) = \mathrm{tr}(1 \cdot \phi W) = \mathrm{tr}(\phi^* 1 \cdot W)$$

for arbitrary $W \in K(H)$. We may thus represent the given effect F by the operator

$$F = \phi^* 1 \tag{2.28}$$

(for which, therefore, we use the same symbol), so that the probability function μ takes the form

$$\mu(F,W) = \mathrm{tr}(FW) \ ,$$

as already mentioned (Eq. (1.16)). Since ϕ^* is real and positive, we have $F = F^* \geq 0$, and $\mathrm{tr}(\phi W) = \mathrm{tr}(FW) \leq 1$ for all $W \in K(H)$ implies $F \leq 1$, so that F indeed belongs to the set $L(H)$ of operators introduced in §1 (Eq. (1.15)).

Every effect which can be measured "non-destructively" by some effect apparatus f, so that this apparatus can be used to perform a selective operation ϕ, is thus described by a unique operator $F \in L(H)$ given by (2.28). As mentioned before, it would not be unreasonable to assume that such f exist in every equivalence class F. This assumption is not even necessary, however, for proving that an arbitrary effect can be represented by an operator $F \in L(H)$, such that the probability function $\mu(F,W)$ takes the form $tr(FW)$. To show this, consider $\mu(F,W)$ with an arbitrary but fixed effect F as a function $\mu_F(W)$ on the set $K(H)$ of density operators W. The physical interpretation of Eq. (1.10) in terms of state mixing implies that μ_F is convex-linear,

$$\mu_F(\lambda W_1 + (1-\lambda)W_2) = \lambda\mu_F(W_1) + (1-\lambda)\mu_F(W_2) \ .$$

Therefore the extension procedure described above for the mapping ϕ may be applied, in exactly the same way, to the function μ_F, leading (first) to a real-linear functional on $B(H)_1^h$ - also denoted by μ_F - which is positive:

$$\mu_F(T) \geq 0 \quad \text{if} \quad T \geq 0 \ . \tag{2.29}$$

(The further extension of μ_F to a complex-linear positive functional on $B(H)_1$ is not needed here.) Since $\mu_F(W) \leq 1$ for all $W \in K(H)$, we obtain, by using (2.22) and (2.23) for an arbitrary $T \in B(H)_1^h$,

$$\begin{aligned} |\mu_F(T)| &= |trT_+ \cdot \mu_F(W_+) - trT_- \cdot \mu_F(W_-)| \\ &\leq trT_+ \cdot \mu_F(W_+) + trT_- \cdot \mu_F(W_-) \\ &\leq trT_+ + trT_- = \|T\|_1 \ ; \end{aligned}$$

i.e., the functional μ_F on $B(H)_1^h$ is trace norm continuous. Therefore it is indeed of the form

$$\mu_F(T) = tr(FT)$$

with a unique $F \in B(H)^h$. Finally, (2.29) implies $F \geq 0$, whereas $F \leq 1$ follows because $tr(FW) = \mu_F(W) = \mu(F,W) \leq 1$ for all W.

Regardless of whether the effect operators F are obtained in this way, or in the form $F = \phi^*1$ as described before, one never finds any conditions for these operators which would go beyond the requirement $F \in L(H)$; in particular, they can not be shown to be projection operators. This already indicates that the whole set L(H) - rather than, e.g., the subset of projection operators - is the "natural" set of effect operators F in quantum mechanics. More arguments supporting this point of view will be presented later.

With $G \in L(H)$, (2.27) implies $\phi^*G \geq 0$ and, as $1 - G \geq 0$, also $\phi^*(1-G) \geq 0$, i.e., $\phi^*G \leq \phi^*1 = F \leq 1$. Thus ϕ^* maps the set L(H) of effects into itself. The effect ϕ^*G has a simple physical interpretation. Imagine that, starting with an ensemble of N systems in some state W, one first performs the selective operation ϕ, corresponding to the effect $F = \phi^*1$, and measures some other effect G afterwards. Let the effects F and G be measured by effect apparatuses f and g, respectively. We then ask: how often are both f and g triggered successively by the same microsystem? Now, f is triggered first by $N_+ = tr(FW)N$ systems, which afterwards are in the new state $\hat{W} = \phi W/tr(FW)$. Therefore, the required number of systems which also trigger the apparatus g is

$$N_{++} = tr(G\hat{W})N_+ = tr(G\cdot\phi W)N = tr(\phi^*G\cdot W)N \ .$$

Thus f and g occur successively in the state W with probability

$$\mu(f,g;W) = N_{++}/N = tr(\phi^*G\cdot W) \ . \qquad (2.30)$$

The effect apparatuses f and g may be considered as parts of a new, composite effect apparatus h, defined to be triggered if both f and g are triggered successively. Then, according to (2.30), the operator describing the corresponding effect (the equivalence class of h) is $H = \phi^*G$. Therefore the effect ϕ^*G describes the successive triggering of the effect apparatuses f and g. As we shall see, different operations ϕ may correspond to the same effect F, which implies that the effect ϕ^*G is not fixed by the effects F and G alone, but also depends on the particular operation ϕ applied - or, in other words, on the particular effect apparatus f used to measure the effect F.

The adjoint ϕ^* of every positive linear mapping ϕ of $B(H)_1$ into itself is normal; i.e., it has the following property (cf. [3], Ch. 2). Consider an increasing sequence ($X_{n+1} \geq X_n$; i.e., $X_{n+1} - X_n \geq 0$) of Hermitean operators $X_n \in B(H)$, $n = 1,2,\ldots$, with $X_n \leq Y = Y^* \in B(H)$ for all n. Such a sequence converges in the ultraweak operator topology to a bounded Hermitean operator $X \leq Y$; i.e., by definition of the ultraweak topology, $tr(X_n T) \underset{n}{\to} tr(XT)$ for all $T \in B(H)_1$ (cf., e.g., [3], Ch. 1). A mapping Ψ of B(H) into itself is called normal if, for every such sequence X_n, ΨX_n also converges ultraweakly to ΨX.

To prove normality of ϕ^*, note that positivity of ϕ^* (Eq. (2.27)) implies $\phi^*X_{n+1} \geq \phi^*X_n$ and $\phi^*X_n \leq \phi^*Y$; therefore ϕ^*X_n has an ultraweak limit $\hat{X}$. This implies, for all $T \in B(H)_1$,

$$tr(\phi^*X_n \cdot T) \underset{n}{\to} tr(\hat{X}T) ,$$

whereas, on the other hand,

$$tr(\phi^*X_n \cdot T) = tr(X_n \cdot \phi T) \underset{n}{\to} tr(X \cdot \phi T) = tr(\phi^*X \cdot T) ,$$

since $X_n \underset{n}{\to} X$ ultraweakly. Thus $tr(\hat{X}T) = tr(\phi^*X \cdot T)$ for all T, which indeed implies $\hat{X} = \phi^*X$.

For physical reasons, the mappings ϕ and ϕ^* must have still another property, called <u>complete positivity</u> and being somewhat stronger than "ordinary" positivity as expressed by (2.11) and (2.27).

Consider a natural number n, an n-dimensional Hilbert space H_n, and the tensor product $H \otimes H_n$ of the state space H, of the quantum mechanical system considered, with H_n. Represent vectors $g \in H_n$ by column vectors,

$$g = \begin{pmatrix} c_1 \\ \cdot \\ \cdot \\ \cdot \\ c_n \end{pmatrix} \equiv (c_i) ,$$

and operators Y on H_n by $n \times n$ matrices,

$$Y = \begin{pmatrix} a_{11} & \cdots & a_{1n} \\ \cdot & & \cdot \\ \cdot & & \cdot \\ \cdot & & \cdot \\ a_{n1} & \cdots & a_{nn} \end{pmatrix} \equiv (a_{ij}) ,$$

with complex numbers c_i and a_{ij}, as usual. Then vectors $\underline{f} \in H \otimes H_n$ may be represented by column vectors with "components" in H,

$$\underline{f} = (f_i) , \quad f_i \in H ,$$

such that, e.g.,

$$(\underline{f},\underline{g}) = \sum_i (f_i,g_i) ,$$

while operators $\underline{X} \in B(H \otimes H_n)$ become $n \times n$ matrices with operator valued "matrix elements",

$$\underline{X} = (X_{ij}) , \quad X_{ij} \in B(H) ,$$

with almost obvious calculation rules like, e.g.,

$$\underline{X}\,\underline{f} = \left(\sum_j X_{ij} f_j\right) .$$

In particular, product vectors $f \otimes g$ with $f \in H$ and $g = (c_i) \in H_n$ are represented by

$$f \otimes g = (c_i f) ,$$

while product operators $X \otimes Y$ with $X \in B(H)$ and $Y = (a_{ij}) \in B(H_n)$ take the form

$$X \otimes Y = (a_{ij}X) .$$

An operator $\underline{T}$ on $H \otimes H_n$ belongs to the trace class,

$$\underline{T} = (T_{ij}) \in B(H \otimes H_n)_1 ,$$

if and only if

$$T_{ij} \in B(H)_1 \quad \text{for all } i,j = 1...n ;$$

i.e., if all "matrix elements" T_{ij} are trace class operators. Any such $\underline{T}$ is thus a finite linear combination of (at most n^2) operators of the form $T \otimes S$ with $T \in B(H)_1$ and $S \in B(H_n)_1 \equiv B(H_n)$. Its trace is given by

$$\mathrm{tr}\underline{T} = \sum_i \mathrm{tr}T_{ii} . \tag{2.31}$$

Now consider an arbitrary (complex-)linear mapping ϕ of $B(H)_1$ into itself, and define a mapping ϕ_n of $B(H \otimes H_n)_1$ into itself - which is also linear - by

$$\phi_n : \underline{T} = (T_{ij}) \to \phi_n\underline{T} = (\phi T_{ij}) . \tag{2.32}$$

Then the original mapping ϕ is called <u>n-positive</u> if ϕ_n is positive; i.e., if

$$\phi_n\underline{T} \geq 0 \quad \text{for} \quad \underline{T} \geq 0 ,$$

and ϕ is called <u>completely positive</u> if it is n-positive for all n. Similarly, a linear mapping Ψ of $B(H)$ into itself yields linear mappings

$$\Psi_n : \underline{X} = (X_{ij}) \to \Psi_n\underline{X} = (\Psi X_{ij}) \tag{2.33}$$

of $B(H \otimes H_n)$ into itself, for any natural number n. Again Ψ is called n-positive if Ψ_n is positive, and Ψ is called completely positive if it is n-positive for all n [6].

If $\Psi : B(H) \to B(H)$ is n-positive, it is also m-positive for all $m < n$. To show this, consider operators $\underline{X} \in B(H \otimes H_n)$ of the particular form

$$\underline{X} = \left(\begin{array}{c|c} \underline{Y} & 0 \\ \hline 0 & 0 \end{array}\right)$$

with arbitrary $m \times m$ operator matrices $\underline{Y}$, representing operators on $B(H \otimes H_m)$, and zeroes everywhere else. Then $\underline{X} \geq 0$ if and only if $\underline{Y} \geq 0$; moreover, we have

$$\Psi_n \underline{X} = \left(\begin{array}{c|c} \Psi_m \underline{Y} & 0 \\ \hline 0 & 0 \end{array}\right) .$$

Thus $\underline{Y} \geq 0$ implies $\underline{X} \geq 0$, so that, if Ψ is n-positive, $\Psi_n \underline{X} \geq 0$, and thus $\Psi_m \underline{Y} \geq 0$; i.e., Ψ is also m-positive. An analogous argument - now involving operator matrices with "matrix elements" from $B(H)_1$ - applies to mappings ϕ of $B(H)_1$ into itself.

Since "ordinary" positivity is the same as 1-positivity, n-positivity with $n > 1$ and, more so, complete positivity, are stronger requirements. A physically relevant example of a positive but not completely positive mapping will be discussed in §3.

A mapping $\phi : B(H)_1 \to B(H)_1$ is n-positive if and only if the adjoint mapping $\phi^* : B(H) \to B(H)$ is n-positive. We show this by first proving that the mapping $(\phi^*)_n$ defined by (2.33) - with ϕ^* for Ψ - is the adjoint of the mapping ϕ_n defined by (2.32); i.e.,

$$(\phi^*)_n = (\phi_n)^* . \tag{2.34}$$

By definition of the adjoint (see Eq. (2.25)), (2.34) means

$$\mathrm{tr}(\underline{X} \cdot \phi_n \underline{T}) = \mathrm{tr}((\phi^*)_n \underline{X} \cdot \underline{T}) \tag{2.35}$$

for arbitrary $\underline{X} = (X_{ij}) \in B(H \otimes H_n)$ and $\underline{T} = (T_{ij}) \in B(H \otimes H_n)_1$. But since, according to "matrix calculus" and Eqs. (2.32) and (2.33), $\underline{X}\cdot\phi_n\underline{T}$ and $(\phi^*)_n\underline{X}\cdot\underline{T}$ have the matrix representations

$$\underline{X}\cdot\phi_n\underline{T} = \Big(\sum_k X_{ik}\cdot\phi T_{kj}\Big) , \qquad (\phi^*)_n\underline{X}\cdot\underline{T} = \Big(\sum_k \phi^* X_{ik}\cdot T_{kj}\Big) ,$$

Eqs. (2.31) and (2.25) indeed lead to (2.35):

$$\begin{aligned} \mathrm{tr}(\underline{X}\cdot\phi_n\underline{T}) &= \sum_{ik} \mathrm{tr}(X_{ik}\cdot\phi T_{ki}) \\ &= \sum_{ik} \mathrm{tr}(\phi^* X_{ik}\cdot T_{ki}) = \mathrm{tr}((\phi^*)_n\underline{X}\cdot\underline{T}) . \end{aligned}$$

According to (2.34), we may simply write ϕ_n^* for $(\phi^*)_n$ or $(\phi_n)^*$.

Now let ϕ_n be positive. Then $\phi_n\underline{W} \geqslant 0$ for all $\underline{W} \in K(H \otimes H_n)$; therefore, with (2.35), $\underline{X} \geqslant 0$ implies

$$\mathrm{tr}(\phi_n^*\underline{X}\cdot W) = \mathrm{tr}(\underline{X}\cdot\phi_n\underline{W}) \geqslant 0$$

for all $\underline{W}$, which in turn implies $\phi_n^*\underline{X} \geqslant 0$, i.e., positivity of ϕ_n^*. Vice versa, let ϕ_n^* be positive. Then $\phi_n^*\underline{W} \geqslant 0$ and thus, again by (2.35),

$$\mathrm{tr}(\underline{W}\cdot\phi_n\underline{T}) = \mathrm{tr}(\phi_n^*\underline{W}\cdot\underline{T}) \geqslant 0$$

for all $\underline{W} \in K(H \otimes H_n)$ if $\underline{T} \in B(H \otimes H_n)_1$, $\underline{T} \geqslant 0$. Therefore $\phi_n\underline{T} \geqslant 0$; i.e., ϕ_n is positive. This proves the above conjecture.

As already mentioned, there are physical reasons for postulating that every operation ϕ - or, equivalently, its adjoint ϕ^* - has to be completely positive [7]. As is well known (and as will be elaborated here in some more detail in §4), the Hilbert space $H \otimes H_n$ can be considered as the state space of a composite system I + II, consisting of the system I considered up to now, with state space H, and another microsystem II with state space H_n. (Since H_n is n-dimensional, system II is usually called an n-level system.) The mappings ϕ_n considered above then acquire a simple physical interpretation.

Assume that there is no interaction between systems I and II. An ensemble of N composite systems then consists of N pairs of non-interacting systems I and II. Its state is described by some density operator $\underline{W}$ on $H \otimes H_n$. Now apply to system I of each pair the selective operation described by the mapping ϕ, while leaving system II unaffected; or, in other words: let system I of each pair interact with a suitable effect apparatus f (which is supposed not to interact with system II), and select those pairs I + II for which the apparatus f is triggered. This procedure clearly defines an operation $\underline{\phi}$ on $K(H \otimes H_n)$, which can be extended to a positive linear mapping of $B(H \otimes H_n)_1$ into itself.

In particular, density operators $\underline{W}$ of the form $\underline{W} = W \otimes V$ with $W \in K(H)$ and $V \in K(H_n)$ describe uncorrelated states of the composite system, which may be prepared by using separate preparing instruments w for system I and v for system II, and combining the single systems I and II into pairs afterwards. (See also §4.) For such states we must have

$$\underline{\phi}(W \otimes V) = \phi W \otimes V ,$$

by the definition of the operation $\underline{\phi}$. On the other hand, since $W \otimes V$ with $V = (v_{ij})$ has the matrix representation $(v_{ij}W)$, an analogous relation,

$$\phi_n(W \otimes V) = \phi W \otimes V ,$$

follows from (2.32) for the mapping ϕ_n. Thus

$$\phi_n(W \otimes V) = \underline{\phi}(W \otimes V)$$

for all $W \in K(H)$ and all $V \in K(H_n)$. Since both ϕ_n and $\underline{\phi}$ are linear, the last equality can be extended, by linearity, first to all operators $W \otimes S$ with $S \in B(H_n)$, then to all $T \otimes S$ with arbitrary $T \in B(H)_1$, and finally to all $\underline{T} \in B(H \otimes H_n)_1$. Thus $\phi_n \equiv \underline{\phi}$; but the latter, as an operation, has to be positive. Therefore ϕ must be n-positive; and n was arbitrary.

Actually the preceding argument could be applied as well with an infinite-dimensional (separable) Hilbert space $\underset{\sim}{H}$ in place of H_n, as describing an "ordinary", rather than an n-level system II. This would lead us to postulate a positivity property for ϕ which appears to be somewhat stronger than complete positivity. We shall prove in §3, however, that this apparently stronger property is already implied by complete positivity.

The effect apparatus f used for the operation ϕ may also be used to perform another selective operation ϕ', called <u>complementary</u> to ϕ. Namely, instead of selecting, after their introduction with f, those systems which have triggered the apparatus f, we may as well select the systems which have <u>not</u> triggered f. By interchanging "yes" and "no" in the verbal interpretation of the response of the effect apparatus f, we get another effect apparatus f', which is triggered if f is not triggered, and vice versa. The corresponding effect operators F and F' are related by

$$F + F' = 1 \ , \qquad (2.36)$$

since the occurence of either F or F' is certain by definition, and therefore

$$\mathrm{tr}(FW) + \mathrm{tr}(F'W) = \mathrm{tr}((F + F')W) = 1 = \mathrm{tr}(1W)$$

for all $W \in K(H)$. ($F \in L(H)$ clearly implies $F' \in L(H)$.) The effect $F' = 1 - F$ describing, in this sense, the non-occuurrence of F, is also called complementary to F - or, briefly, "not F". By definition, ϕ and F are also complementary to ϕ' and F', respectively:

$$(\phi')' = \phi \ , \quad (F')' = F \ ;$$

complementarity is thus a symmetric relation. Since the operation ϕ' is performed by selecting according to the occurence of F', two operations ϕ and ϕ' complementary to each other are characterized by the complementarity (2.36) of the corresponding effects $F = \phi^*1$ and $F' = \phi'^*1$ or, equivalently, by

$$\mathrm{tr}(\phi W) + \mathrm{tr}(\phi' W) = 1 \quad \text{for all } W \in K(H) \ . \qquad (2.37)$$

Whereas, by (2.36), an effect F has exactly one complementary effect F', there are in general many different operations ϕ' complementary to a given operation ϕ, since (2.37) with given ϕ does not uniquely determine ϕ'. Both ϕ and ϕ' thus depend on the particular apparatus f used to measure the effect F, rather than on F only.

Finally, as already mentioned at the beginning of this chapter, the effect apparatus f may also be used to perform a <u>non-selective</u> operation. In this case, no selection with respect to the response of the apparatus f is made - or, equivalently, the two (by definition disjoint) subensembles of systems produced by the complementary operations ϕ and ϕ' are mixed afterwards. Thereby an ensemble of N systems in an initial state W is transformed into another ensemble, consisting again of N systems, of which $N_+ = \mathrm{tr}(\phi W)\cdot N$ are in the state $\hat{W} = \phi W/\mathrm{tr}(\phi W)$, whereas the remaining $N_- = \mathrm{tr}(\phi' W)\cdot N$ systems are in the state $\hat{W}' = \phi' W/\mathrm{tr}(\phi' W)$, according to the physical meaning of ϕ (Eqs. (2.5), (2.6)) and ϕ'. The final state $\tilde{W}$ resulting from the non-selective operation is thus a mixture of the states $\hat{W}$ and $\hat{W}'$ with the weights $N_+/N = \mathrm{tr}(\phi W)$ and $N_-/N = \mathrm{tr}(\phi' W)$, respectively; i.e.,

$$\tilde{W} = \mathrm{tr}(\phi W)\hat{W} + \mathrm{tr}(\phi' W)\hat{W}' = \phi W + \phi' W \ .$$

Therefore the mapping $\tilde{\phi} : K(H) \rightarrow K(H)$ describing the non-selective operation (Eq. (2.1)) is simply given by

$$\tilde{\phi} = \phi + \phi' : W \rightarrow \tilde{W} = \phi W + \phi' W \; . \tag{2.38}$$

(Since ϕ and ϕ' are positive, we have $\tilde{W} \geq 0$, while (2.37) implies $\mathrm{tr}\tilde{W} = 1$. Thus $\tilde{\phi}$ indeed maps $K(H)$ into itself.) Like ϕ and ϕ', $\tilde{\phi}$ may also be extended to a mapping of $B(H)_1$ into itself. As is obvious from (2.38), this extension is given by

$$\tilde{\phi} = \phi + \phi' : T \rightarrow \tilde{T} = \phi T + \phi' T \tag{2.39}$$

for arbitrary $T \in B(H)_1$.

Since the adjoint $\tilde{\phi}^*$ of $\tilde{\phi}$ is $\phi^* + \phi'^*$, the "effect" corresponding to the nonselective operation $\tilde{\phi}$ is

$$\tilde{F} = \tilde{\phi}^* 1 = \phi^* 1 + \phi'^* 1 = F + F' = 1 \; . \tag{2.40}$$

This is also obvious from (and equivalent to)

$$\mathrm{tr}(\tilde{F}W) \equiv \mathrm{tr}(\tilde{\phi}W) = 1 \quad \text{for all } W \in K(H) \; ,$$

which expresses the fact that the transition probability is identically one, and is thus characteristic of a non-selective operation. The "effect" $\tilde{F} = 1$ describes the trivial "yes-no measurement" which always gives the result "yes". An effect apparatus $\tilde{f}$ measuring this effect thus simply counts the systems in any given ensemble. As non-selective operations can thus be considered formally as very particular "selective" ones - the "selection" being made with respect to the trivial effect $\tilde{F} = 1$ - they need no separate mathematical treatment.

These considerations are illustrated by the example of "ideal measurements" (Eqs. (2.8) and (2.9)). In this case, the operation ϕ' complementary to ϕ is given by $\phi' W = E'WE'$.

Summarizing the main content of this chapter, we may state that every operation has to be described mathematically by a completely positive complex-linear mapping ϕ of $B(H)_1$ into itself, which satisfies $tr(\phi W) \leq 1$ for all $W \in K(H)$. In the absence of any additional selection criteria for "physically realizable" operations, we shall assume in the following that also, vice versa, every mapping ϕ with these properties describes an operation. Instead of defending this assumption by more or less sophisticated arguments, we only remind the reader that similar assumptions are quite usual in quantum mechanics. For instance, one almost always assumes that every projection operator E on H describes a yes-no measurement which could be performed, at least "in principle", by a "suitable" apparatus, even if nobody knows - except in a few particular cases corresponding, e.g., to position measurements - what such an apparatus would look like in practice.

§3 The First Representation Theorem

An arbitrary effect can be represented, as shown in the preceding §2, by an operator F on the state space H of the system considered. We shall now derive a corresponding explicit representation of an arbitrary operation ϕ in terms of operators on H. This representation is provided by

Theorem 1 (First Representation Theorem):

For an arbitrary operation ϕ, there exist operators A_k, $k \in K$ (a finite or countably infinite index set) on the state space H, satisfying

$$\sum_{k \in K_o} A_k^* A_k \leq 1 \quad \text{for all finite subsets } K_o \subset K \ , \tag{3.1}$$

such that, with arbitrary $T \in B(H)_1$ and $X \in B(H)$, the mappings ϕ and ϕ^* are given by

$$\phi T = \sum_{k \in K} A_k \, T \, A_k^* \tag{3.2}$$

and

$$\phi^* X = \sum_{k \in K} A_k^* \, X \, A_k \ , \tag{3.3}$$

respectively. In particular, the effect F corresponding to ϕ is given by

$$F = \phi^* 1 = \sum_{k \in K} A_k^* A_k \ . \tag{3.4}$$

If the index set K is infinite, (3.1) implies that - independently of the ordering of K - the infinite sum in (3.2) converges in the trace norm topology, while the infinite sums in (3.3) and (3.4) converge ultraweakly, thus defining the precise meaning of these equations.

Vice versa, given any countably or even uncountably infinite set of operators A_k on H, $k \in K$, satisfying condition (3.1), then (3.2) defines an operation ϕ, whose adjoint ϕ^* and corresponding effect F are given by (3.3) and (3.4), respectively.-

Proof: We first discuss some of the more technical details. First, (3.1) implies that A_k can be different from zero for at most countably many indices $k \in K$; therefore, if in the second part of the theorem an uncountably infinite K occurs, it may be replaced by an at most countably infinite subset, since operators $A_k = 0$ do not contribute in Eqs. (3.2) to (3.4). To prove the above statement, consider a countable dense set of vectors $f_i \in H$, $i = 1,2,\ldots$. Keeping i fixed, and choosing an arbitrary natural number n, (3.1) implies that

$$(f_i, A_k^* A_k f_i) = \|A_k f_i\|^2 > 1/n$$

may be true for finitely many indices k only. Therefore the subset $K_i \subseteq K$ defined by

$$\|A_k f_i\| \neq 0 \quad \text{if} \quad k \in K_i \tag{3.5}$$

is at most countably infinite. The same then is true for the union $\cup_i K_i$. By (3.5), $A_k f_i = 0$ for all i if $k \notin \cup_i K_i$. But (3.1) also implies $A_k^* A_k \leq 1$, and thus $\|A_k\| \leq 1$, for all k. Therefore all A_k are bounded, and thus $A_k = 0$ for all $k \notin \cup_i K_i$, by continuity.

Assume therefore, in the following, that K is countably infinite, and can thus be identified, after a suitable ordering, with the set of natural numbers: $K = \{1,2,\ldots\}$. The somewhat simpler case of a finite set $K = \{1,\ldots,N\}$ need not be treated separately, since we may formally enlarge K by setting $A_k = 0$ for $k > N$.

Consider the operators

$$F_n \underset{\text{df.}}{=} \sum_{k \leq n} A_k^* A_k \,.$$

Since $A_k^* A_k \geq 0$, these operators are non-negative and increasing with n, and satisfy $F_n \leq 1$ for all n by (3.1). Therefore they converge ultraweakly, for $n \to \infty$, to an operator

$$F = \sum_n A_k^* A_k$$

- which defines the meaning of the infinite sum in (3.4) - with $0 \leq F \leq 1$; i.e., $F \in L(H)$. More generally, take an arbitrary $X \geq 0$ from $B(H)$, so that $0 \leq X \leq \|X\| \cdot 1$. Since $A_k^* X A_k \geq 0$ and $A_k^*(\|X\| \cdot 1 - X)A_k \geq 0$ for all k, the operators

$$\hat{X}_n \overset{\text{df.}}{=} \sum_{k \leq n} A_k^* X A_k$$

increase with n, and satisfy $0 \leq \hat{X}_n \leq \|X\| \cdot F_n \leq \|X\| \cdot F$ for all n. Thus the sequence $\hat{X}_n$, $n = 1,2,\ldots$ also converges ultraweakly to some operator

$$\hat{X} = \sum_k A_k^* X A_k \geq 0 .$$

Since $\hat{X}_n$ depends linearly on X, and any $X \in B(H)$ is a linear combination of at most four non-negative operators, $\hat{X}_n$ converges ultraweakly, thus giving a meaning to Eq. (3.3), for an arbitrary $X \in B(H)$. Moreover, these ultraweak limits defining the right hand sides of (3.3) and (3.4) do not depend on the particular ordering of the index set K. For suppose $K = \{1,2,\ldots\}$ to be reordered as, e.g., $K = \{k_1,k_2,\ldots\}$. An arbitrary $X \geq 0$ now leads to a sequence of operators

$$\tilde{X}_n = \sum_{i \leq n} A_{k_i}^* X A_{k_i}$$

which, as above, converges ultraweakly to an operator $\tilde{X}$. Since $\{k_1,k_2,\ldots\}$ is a reordering of $\{1,2,\ldots\}$, there are suitable numbers m and m' for each n, such that $\tilde{X}_n \leq \hat{X}_m \leq \hat{X}$ and $\hat{X}_n \leq \tilde{X}_{m'} \leq \tilde{X}$. Therefore also $\tilde{X} \leq \hat{X}$ and $\hat{X} \leq \tilde{X}$, which implies $\tilde{X} = \hat{X}$. By linearity, the same holds true for an arbitrary $X \in B(H)$, which proves the order independence of the right hand side of (3.3), of which (3.4) is merely a particular case.

Since the trace class $B(H)_1$ is a two-sided ideal in $B(H)$ (cf. [3], Ch. 1, or [4]), so that $XT \in B(H)_1$ and $TX \in B(H)_1$ for arbitrary

$T \in B(H)_1$ and $X \in B(H)$, we have

$$\hat{T}_n \underset{\mathrm{df.}}{\equiv} \sum_{k<n} A_k\, T\, A_k^* \quad \in B(H)_1^+$$

for all $T \in B(H)_1^+$. (Note that, obviously, $A_k T A_k^* \geq 0$ if $T \geq 0$.) Now we have

$$\|\hat{T}_m - \hat{T}_n\|_1 = \| \sum_{k=n+1}^{m} A_k\, T\, A_k^* \|_1 = \mathrm{tr}(\sum_{k=n+1}^{m} A_k\, T\, A_k^*)$$

$$= \mathrm{tr}(\sum_{k=n+1}^{m} A_k^*\, A_k\, T) = \mathrm{tr}(F_m T) - \mathrm{tr}(F_n T) \;,$$

due to the cylic interchangeability of operators under the trace, whereas, on the other hand,

$$\mathrm{tr}(F_n T) \underset{n}{\to} \mathrm{tr}(FT) \;,$$

by the ultraweak convergence $F_n \underset{n}{\to} F$. Therefore the operators $\hat{T}_n$ form a Cauchy sequence with respect to the trace norm, so that there exists a $\hat{T} \in B(H)_1$ with $\|\hat{T}_n - \hat{T}\|_1 \underset{n}{\to} 0$, which gives a meaning to the right hand side of Eq. (3.2):

$$\hat{T} \underset{\mathrm{df.}}{\equiv} \sum_k A_k\, T\, A_k^* \;.$$

Since trace norm convergence implies convergence in the operator norm (uniform) topology, and therefore also in the ultraweak topology, $\hat{T}$ is also the ultraweak limit of the operators $\hat{T}_n$ which, moreover, are non-negative and increasing with n. This yields $T \geq 0$ and the independence of $\hat{T}$ on the order of summation, as above. As arbitrary trace class operators are linear combinations of nonnegative ones, the right hand side of (3.2) exists, and is order-independent, for arbitrary $T \in B(H)_1$ also. Infinite sums of the type (3.2), (3.3) and (3.4) shall be understood, from now on, to represent limits of finite sums in the appropriate operator topologies.

We are now ready to prove the second part of Theorem 1. Assume operators A_k, $k \in K$ satisfying (3.1) to be given. Then the mappings ϕ and ϕ^* given by (3.2) and (3.3) are well-defined and positive, as just proved; obviously, they are also linear. It remains to show that ϕ^* is the adjoint of ϕ, and that ϕ is completely positive.

For arbitrary $X \in B(H)$ and $T \in B(H)_1$, we have

$$\begin{aligned} tr(X \cdot \hat{T}_n) &= tr(X \cdot \sum_{k \leq n} A_k \; T \; A_k^*) \\ &= tr(\sum_{k \leq n} A_k^* \; X \; A_k \cdot T) = tr(\hat{X}_n \cdot T) \; . \end{aligned}$$

Since

$$\begin{aligned} tr(X\hat{T}_n) - tr(X\hat{T}) &= tr(X(\hat{T}_n - \hat{T})) \\ &\leq \|X\| \; \|\hat{T}_n - \hat{T}\|_1 \underset{n}{\rightarrow} 0 \; , \end{aligned}$$

$tr(X\hat{T}_n)$ converges to $tr(X\hat{T}) = tr(X \cdot \phi T)$ whereas, since

$$\hat{X}_n \underset{n}{\rightarrow} \hat{X} = \phi^* X$$

ultraweakly, $tr(\hat{X}_n T)$ converges to $tr(\phi^* X \cdot T)$. Thus ϕ^* is indeed the adjoint of ϕ.

To prove complete positivity, consider an arbitrary finite-dimensional or separable Hilbert space $\underset{\sim}{H}$, and define a mapping $\underline{\phi}$ of $B(H \otimes \underset{\sim}{H})_1$ into itself by

$$\underline{\phi} \; \underline{T} = \sum_{k \in K} (A_k \otimes \underset{\sim}{1}) \; \underline{T} \; (A_k \otimes \underset{\sim}{1})^* \tag{3.6}$$

for arbitrary $\underline{T} \in B(H \otimes \tilde{H})_1$. Since

$$\sum_{k \in K_o} (A_k \otimes \underset{\sim}{1})^* \; (A_k \otimes \underset{\sim}{1}) = (\sum_{k \in K_o} A_k^* \; A_k) \otimes \underset{\sim}{1} \leq \underline{1}$$

for any finite subset $K_o \subseteq K$ (with $\underline{1}$ denoting the unit operator on

$H \otimes \underset{\sim}{H})$, $\underline{\phi}$ is a mapping of the form (3.2) with the operators A_k replaced by $A_k \otimes \underset{\sim}{1}$, which also satisfy a condition of the form (3.1). Therefore $\underline{\phi}$ is well-defined, linear, and positive. If, in particular, $T = T \otimes S$ with $T \in B(H)_1$ and $S \in B(\underset{\sim}{H})_1$, (3.6) yields

$$\underline{\phi}(T \otimes S) = \sum_{k\in K} (A_k \otimes \underset{\sim}{1})(T \otimes S)(A_k \otimes \underset{\sim}{1})^*$$

$$= (\sum_{k\in K} A_k \; T \; A_k^*) \otimes S = \phi T \otimes S \; .$$

Taking for $\underset{\sim}{H}$ an n-dimensional Hilbert space H_n, the last equation means that $\underline{\phi}$ coincides on operators of the form $T \otimes S$ - and therefore, by linearity, on the whole space $B(H \otimes H_n)_1$ - with the mapping ϕ_n introduced in §2 to define n-positivity of ϕ. Since $\underline{\phi} = \phi_n$ is positive, ϕ is n-positive, and thus completely positive, since n was arbitrary.

Actually the above reasoning proves even more. First, since $\underline{\phi}$ is again of the form (3.2), it is also completely positive. Therefore the mappings ϕ_n introduced in §2 share this property, and can indeed be interpreted physically as operations. Second, since $\underset{\sim}{H}$ may also be infinite-dimensional, mappings ϕ of the form (3.2) have a positivity property which at first sight looks somewhat stronger than complete positivity. As was already mentioned in §2, one might have been tempted to include this positivity property in the very definition of an operation ϕ by postulating the existence of positive linear mappings $\underline{\phi}$ of $B(H \otimes \underset{\sim}{H})_1$, which satisfy

$$\underline{\phi}(W \otimes \underset{\sim}{W}) = \phi W \otimes \underset{\sim}{W} \tag{3.7}$$

for uncorrelated states $W \otimes \underset{\sim}{W}$, for all finite- or infinite-dimensional Hilbert spaces $\underset{\sim}{H}$. (The effects $\underline{F} = \underline{\phi}^* \underline{1}$ corresponding to $\underline{\phi}$ satisfy

$$\mathrm{tr}(\underline{F}(W \otimes \underset{\sim}{W})) = \mathrm{tr}(\underline{\phi}(W \otimes \underset{\sim}{W})) = \mathrm{tr}(\phi W \otimes \underset{\sim}{W})$$

$$= \mathrm{tr}(\phi W) = \mathrm{tr}(FW) = \mathrm{tr}((F \otimes \underset{\sim}{1})(W \otimes \underset{\sim}{W})) \; ,$$

which implies - for details see §4 -

$$\underline{F} = F \otimes \underset{\sim}{1} \, . \tag{3.8}$$

The same follows from the explicit form (3.6) of $\underline{\phi}$, which yields

$$\underline{F} = \underline{\phi}^* \underline{1} = \sum_{k \in K} (A_k \otimes \underset{\sim}{1})^* (A_k \otimes \underset{\sim}{1}) = (\sum_{k \in K} A_k^* A_k) \otimes \underset{\sim}{1} = F \otimes \underset{\sim}{1}$$

in analogy to (3.4).) Since any finite-dimensional H_n is contained in a given infinite-dimensional $\underset{\sim}{H}$, and thus $B(H \otimes H_n)_1$ is embedded in $B(H \otimes \underset{\sim}{H})_1$ (as explained in more detail in §2 for the case of finite dimensions of both H_n and $\underset{\sim}{H}$), it would also have been sufficient to postulate (3.7) for infinite-dimenisonal $\underset{\sim}{H}$ only. However, the proof we are about the give of the first part of Theorem 1 demonstrates that complete positivity of ϕ is already sufficient to derive Eq. (3.2) for ϕ, which then, as has just been proved, also implies the stronger positivity property. Moreover, in contrast to the latter, complete positivity has already been studied by mathematicians, and is now also a standard postulate for operations and related state maps in the physical literature (cf., e.g., [7], [3], Ch. 9, [8] and [9]).

The basic ingredient for the proof of the remaining part of Theorem 1 is

<u>Stinespring's Theorem:</u> A complex-linear mapping Ψ of a concrete C*-algebra $\mathbf{B} \subseteq B(H)$ into $B(H)$ is completely positive if and only if it is of the form

$$\Psi : X \in \mathbf{B} \rightarrow \Psi X = \mathbf{A}^* \pi(X) \mathbf{A}$$

with a representation π of $\mathbf{B}$ on a Hilbert space $\tilde{H}$, and a bounded operator $\mathbf{A}$ mapping H into $\tilde{H}$.

Rather than proving this here (for proofs, see the original paper [6] or [3], Ch. 9), we shall only explain the basic notions. A concrete C*-algebra $\mathbf{B}$ is a C*-algebra of operators on a Hilbert space H, i.e., a subset of $B(H)$ which is closed under multiplication, Hermitean conjugation, and complex linear combination of operators, as well as with respect to the uniform (operator norm) topology, and which contains the unit operator 1. In particular, B(H) itself is a concrete C*-algebra. A representation π of $\mathbf{B}$ on a Hilbert space $\tilde{H}$ is a complex-linear mapping

$$\pi : X \in \mathbf{B} \rightarrow \pi(X) \in B(\tilde{H})$$

of $\mathbf{B}$ into $B(\tilde{H})$ which preserves operator products and satisfies $\pi(X^*) = (\pi(X))^*$ and $\pi(1) = \tilde{1}$ (the unit operator on $\tilde{H}$). (The image $\pi(\mathbf{B}) \subseteq B(\tilde{H})$ of $\mathbf{B}$ under π is then also a concrete C*-algebra).

We shall, moreover, also use the following lemmas:

<u>Lemma 1:</u> Let π be a representation of B(H) on a Hilbert space $\tilde{H}$. Then there are mutually orthogonal subspaces $H_k \subseteq \tilde{H}$, $k \in K$ (a finite - sometimes even empty - or countably or uncountably infinite index set), which, together with the orthogonal complement H_o in $\tilde{H}$ of the direkt sum $\bigoplus_{k\in K} H_k$ of these subspaces H_k, are invariant under the operators $\pi(X)$ for all $X \in B(H)$. All subspaces H_k can be identified with H in such a way that $\pi(X)$, when restricted to any such $H_k \equiv H$, acts like the identical representation; i.e., $\pi(X) = X$ on H_k. The restriction $\pi_o(X)$ of $\pi(X)$ to H_o satisfies $\pi_o(E) = 0$ for all finite-dimensional projection operators E. In other words, the orthogonal decomposition

$$\tilde{H} = H_o \oplus (\bigoplus_{k\in K} H_k), \quad H_k \equiv H$$

of $\tilde{H}$ induces the reduction

$$\pi = \pi_o \oplus (\bigoplus_{k\in K} \pi_k), \quad \pi_k(X) \equiv X$$

of the representation π. The subrepresentation π_o is absent if H_o consists of the zero vector only. (Technically speaking, π_o is a representation of the quotient algebra B(H)/C(H) of B(H) with respect to the norm-closed two-sided ideal C(H) of all compact operators on H.)

For a proof see, e.g., [10]. (Compare also [3], Ch. 9.)

Lemma 2: Let H_1 be a separable Hilbert space, and A a bounded linear operator from H_1 into a Hilbert space H_2 which need not be separable. Then $T \in B(H_1)_1$ implies $ATA^* \in B(H_2)_1$, and

$$tr_2(X \cdot ATA^*) = tr_1(A^*XA \cdot T)$$

for all $X \in B(H_2)$. (Here the traces in H_1 and H_2 are distinguished, for clarity, by corresponding suffices. If $H_1 = H_2$, Lemma 2 reduces to well-known facts.)

Proof of Lemma 2: Since A is bounded, it maps a dense set of vectors in H_1 into a set of vectors in H_2 which is dense in the range AH_1 of A. As H_1 is separable, it thus follows that AH_1 is contained in a separable subspace H_2' of H_2; i.e., $A = E_2A$ with the projection operator E_2 onto H_2'. The adjoint A^* of A - a bounded operator from H_2 into H_1 defined by $(f_1, A^*f_2) = (Af_1, f_2)$ for arbitrary $f_1 \in H_1$ and $f_2 \in H_2$ - then satisfies $A^* = A^*E_2$.

Since both H_1 and H_2' are separable, there is an isometric operator $U : H_1 \to H_2$ with range $UH_1 = H_2'$; i.e., $U^*U = 1_1$ (the unit operator on H_1), and $UU^* = E_2$. Both U^*A and A^*U, then, are bounded operators on H_1, so that $T \in B(H_1)_1$ implies $U^*ATA^*U \in B(H_1)_1$. Since U maps H_1 isomorphically into H_2', the last relation implies that $U(U^*ATA^*U)U^* = E_2ATA^*E_2 = ATA^*$ is a trace class operator on H_2'. Being zero, moreover, on the orthogonal complement of H_2', this operator indeed belongs to $B(H_2)_1$.

With $X \in B(H_2)$ arbitrary, then, $X \cdot ATA^*$ also belongs to $B(H_2)_1$. Since this operator, too, is zero on the orthogonal complement of H_2', its trace can be evaluated with an orthogonal basis $\{f_2^i\}$ of the subspace H_2' of H_2. We then obtain

$$\mathrm{tr}_2(X\cdot ATA^*) = \sum_i (f_2^i, XATA^*f_2^i) = \sum_i (UU^*f_2^i, XATA^*UU^*f_2^i)$$

$$= \sum_i (U^*f_2^i, U^*XATA^*U(U^*f_2^i)) = \mathrm{tr}_1(U^*XATA^*U) ,$$

because $UU^*f_2^i = E_2f_2^i = f_2^i$, and $\{U^*F_2^i\}$ is an orthogonal basis in H_1. Both U^*XA and A^*U belong to $B(H_1)$. Therefore U^*XAT belongs to $B(H_1)_1$, and may be interchanged with A^*U under the trace, which finally leads to

$$\mathrm{tr}_1(A^*UU^*XAT) = \mathrm{tr}_1(A^*E_2XAT) = \mathrm{tr}_1(A^*XA\cdot T) ,$$

q.e.d.-

After these preparations, we can now also prove the first part of Theorem 1. For this it suffices to show that ϕ^* is of the form (3.3), with suitable operators A_k ($k \in K$) satisfying (3.1), since this immediately implies Eq. (3.2) for ϕ. Indeed, once the representation (3.3) for ϕ^* is proved, we know from the previous discussion that ϕ^*, besides being the adjoint of the operation ϕ, is also the adjoint of the mapping defined by Eq. (3.2). By the definition (2.25) of the adjoint, however, two mappings of $B(H)_1$ into itself with the same adjoint must be identical.

The mapping ϕ^* satisfies the assumption of Stinespring's Theorem with $\mathbf{B} = B(H)$. Thus, for all $X \in B(H)$,

$$\phi^*X = \mathbf{A}^* \pi(X) \mathbf{A} \tag{3.9}$$

with a bounded operator $\mathbf{A} : H \to \tilde{H}$ and a representation π of $B(H)$ on $\tilde{H}$. According to Lemma 1, π is reduced by a direct sum decomposition of $\tilde{H}$ into subspaces H_o and H_k, $k \in K$. Denote by E_o and E_k the projection operators on $\tilde{H}$ onto these subspaces H_o and H_k, respectively, and introduce the operators

$$\tilde{A}_o = E_o \mathbf{A} , \quad \tilde{A}_k = E_k \mathbf{A} \quad (k \in K) \tag{3.10}$$

mapping H into $\tilde{H}$, whose adjoints, mapping $\tilde{H}$ into H, are given by

$$\tilde{A}^*_o = \mathbf{A}^* E_o , \quad \tilde{A}^*_k = \mathbf{A}^* E_k \tag{3.11}$$

with the adjoint $\mathbf{A}^* : \tilde{H} \to H$ of $\mathbf{A}$. Since the ranges of $\tilde{A}_o$ and $\tilde{A}_k$ are contained in H_o and $H_k \equiv H$, respectively, the operators (3.10) may as well be reinterpreted as mapping H into H_o or H, respectively. If interpreted in this way, we denote them by A_o and A_k, respectively. Then we obtain

$$A^*_o = \mathbf{A}^*|_{H_o} , \quad A^*_k = \mathbf{A}^*|_{H_k} ; \tag{3.12}$$

i.e., A^*_o and A^*_k are the restrictions of $\mathbf{A}^*$ to H_o and H_k, respectively. Indeed, A^*_o as given by (3.12) satisfies

$$(A^*_o f_o, f) = (\mathbf{A}^* f_o, f) = (f_o, \mathbf{A} f) = (E_o f_o, \mathbf{A} f)$$

$$= (f_o, E_o \mathbf{A} f) = (f_o, \tilde{A}_o f) = (f_o, A_o f)$$

for all $f_o \in H_o$ and $f \in H$, and an analogous relation follows for A_k. (Remember that $(\mathbf{A}^* \tilde{g}, f) = (\tilde{g}, \mathbf{A} f)$ for arbitrary $\tilde{g} \in \tilde{H}$ and $f \in H$.)

Denoting by K_o an arbitrary finite subset of the index set K, we define a corresponding projection operator on $\tilde{H}$ by

$$E_{K_o} = \sum_{k \in K_o} E_k .$$

Since

$$E_{K_o} \leq \tilde{1} = \pi(1) ,$$

we get with (3.9)

$$\mathbf{A}^* E_{K_o} \mathbf{A} \leq \mathbf{A}^* \mathbf{A} = \mathbf{A}^* \pi(1) \mathbf{A} = \phi^* 1 = F \leq 1 .$$

On the other hand, we have

$$\mathbf{A}^* E_{K_o} \mathbf{A} = \sum_{k\varepsilon K_o} \mathbf{A}^* E_k \mathbf{A} = \sum_{k\varepsilon K_o} \tilde{A}_k^* \tilde{A}_k = \sum_{k\varepsilon K_o} A_k^* A_k ,$$

as follows immediately from the above definitions of $\tilde{A}_k$ and A_k. Therefore the operators A_k on H defined above satisfy the condition (3.1) of the theorem.

But then, as has been shown before, the index set K' defined by $A_k \neq 0$ iff $k \varepsilon K'$ is at most countably infinite. By definition, $A_k = 0$ implies $\tilde{A}_k = E_k \mathbf{A} = 0$. Therefore, with

$$E' = E_o + \sum_{k\varepsilon K'} E_k = \tilde{1} - \sum_{k\notin K'} E_k ,$$

we have

$$(\tilde{1} - E')\mathbf{A} = \sum_{k\notin K'} E_k \mathbf{A} = \sum_{k\notin K'} \tilde{A}_k = 0 ;$$

i.e., $\mathbf{A} = E'\mathbf{A}$. In (3.9) we may thus replace $\mathbf{A}$ by

$$E'\mathbf{A} : H \to E'\tilde{H} \underset{df.}{\equiv} \tilde{H}'$$

and $\mathbf{A}^*$ by $(E'\mathbf{A})^* = \mathbf{A}^* E' : \tilde{H}' \to H$, so that only the subrepresentation $\pi|_{\tilde{H}'}$ of π, acting as

$$\pi_o \oplus (\bigoplus_{k\varepsilon K'} \pi_k), \quad \pi_k(X) \equiv X \tag{3.13}$$

on

$$\tilde{H}' = H_o \oplus (\bigoplus_{k\varepsilon K'} H_k), \quad H_k \equiv H$$

actually enters Eq. (3.9), whereas the orthogonal complement of $\tilde{H}'$ may be completely ignored. Or, in other words: the index set K occurring in Lemma 1 may be identified with K', and thus assumed to be at most countably infinite, without loss of generality.

Considering first the case of an infinite index set K, we may thus simply take $K = \{1,2,...\}$. Define projection operators

$$E_{(n)} = E_o + \sum_{k<n} E_k .$$

Then, using the definition of the operators A_o and A_k and the decomposition (3.13) of π, we obtain

$$\mathbf{A}^* E_{(n)}\pi(X) E_{(n)}\mathbf{A} = A_o^* \pi_o(X) A_o + \sum_{k<n} A_k^* X A_k \tag{3.14}$$

for an arbitrary $X \varepsilon B(H)$. As has already been proved, the last term on the right hand side converges ultraweakly to

$$\sum_k A_k^* X A_k$$

for $n \to \infty$, since condition (3.1) is satisfied. We want to show now that for $n \to \infty$ the left hand side of (3.14) converges ultraweakly to $\mathbf{A}^* \pi(X) \mathbf{A} = \phi^*X$.

By definition, $E_{(n)}$ commutes with $\pi(X)$ for arbitrary $X \varepsilon B(H)$ and all n. This implies

$$E_{(n)}\pi(X) E_{(n)} = E_{(n)}\pi(X) \underset{n}{\to} \pi(X) \text{ ultraweakly,} \tag{3.15}$$

since $E_{(n)} \underset{n}{\to} \tilde{I}$ ultraweakly. (The last statement, being equivalent to

$$tr(E_{(n)}\tilde{T}) \underset{n}{\to} tr\,\tilde{T}$$

for arbitrary $\tilde{T} \varepsilon B(\tilde{H})_1$, immediately follows from the definition of the trace.) Applying Lemma 2 with $H_1 = H$, $H_2 = \tilde{H}$, $A = \mathbf{A}$, an arbitrary $T \varepsilon B(H)_1$, and $E_{(n)}\pi(X)E_{(n)} \varepsilon B(\tilde{H})$ in place of X, we obtain

$$tr_H(\mathbf{A}^* E_{(n)}\pi(X) \mathbf{A}\, T) = tr_{\tilde{H}}(E_{(n)}\pi(X)E_{(n)} \mathbf{A}\, T\, \mathbf{A}^*) .$$

By (3.15), the right hand side converges for $n \to \infty$ to

$$tr_{\tilde{H}}(\pi(X) \mathbf{A}\, T\, \mathbf{A}^*) = tr_H(\mathbf{A}^* \pi(X) \mathbf{A}\, T) ,$$

the last equality following again from Lemma 2, now with $\pi(X)$ for X. As $T \varepsilon B(H)_1$ was arbitrary, this shows that

$$\mathbf{A}^* E_{(n)}\pi(X) E_{(n)}\mathbf{A} \underset{n}{\to} \mathbf{A}^* \pi(X) \mathbf{A} \text{ ultraweakly,} \tag{3.16}$$

as announced above.

Taking thus the ultraweak limit of Eq. (3.14), we obtain

$$\phi^*X = \mathbf{A}^* \, \pi(X) \, \mathbf{A} = A_o^* \, \pi_o(X) \, A_o + \sum_{k\varepsilon K} A_k^* \, X \, A_k \, . \qquad (3.17)$$

If K is finite, the argument leading to (3.14) yields (3.17) directly.

The second summand on the right hand side of (3.17) is already of the form (3.3). Therefore, being the adjoint of a corresponding mapping of $B(H)_1$ into itself as given by (3.2), this term defines a normal mapping of B(H) into itself. But ϕ^*, as the adjoint of the operation ϕ, is also normal; thus the same must be true for the mapping

$$\phi_o : X \to A_o^* \, \pi_o(X) \, A_o \, .$$

But this implies $\phi_o X = 0$ for all $X \varepsilon B(H)$: First, every projection operator E is the (ultraweak) upper limit of finite-dimensional projection operators E_n; but $\pi_o(E_n) = 0$ by Lemma 1, and thus $\phi_o E = 0$. Second, by exploiting the spectral theorem, every Hermitean $X \varepsilon B(H)$ may be represented as an upper limit of operators X_n which are finite linear combinations of projection operators. This implies $\phi_o X = 0$ for Hermitean X and thus, by linearity, for all X. Therefore, finally, Eq. (3.17) reduces to

$$\phi^*X = \sum_{k\varepsilon K} A_k^* \, X \, A_k \, ,$$

which is the required representation (3.3), and Theorem 1 is proved.-

A slight but useful generalization of Theorem 1 is immediately obvious. Consider, together with ϕ, an operation ϕ' complementary to ϕ, and the corresponding non-selective operation $\tilde{\phi} = \phi + \phi'$. Since ϕ' is also an operation, there exists another finite or countably infinite set of operators A_k', $k \varepsilon K'$, so that obvious analogs of Eqs. (3.1) to (3.4) hold true. Moreover, the condition $F + F' = 1$ implies

$$\sum_{k\varepsilon K} A_k^* \, A_k + \sum_{k\varepsilon K'} A_k'^* \, A_k' = 1 \, .$$

A simplified notation is obtained by chosing the index sets K and K' to be disjoint, with $K \cup K' = \{1,\dots,N\}$ or $\{1,2,\dots\}$. This allows us to drop the primes on the operators A'_k with $k \in K'$, and we arrive at

Theorem 1': Given any two complementary operations ϕ and ϕ', there exist an index set $J = \{1,\dots,N\}$ or $\{1,2,\dots\}$, operators A_k, $k \in J$, with

$$\sum_{k\in J} A_k^* A_k = 1 \;, \tag{3.18}$$

and two complementary subsets K and K' of J, such that, for all $T \in B(H)_1$ and $X \in B(H)$,

$$\phi T = \sum_{k\in K} A_k T A_k^* \;, \qquad \phi^* X = \sum_{k\in K} A_k^* X A_k \;, \tag{3.19}$$

and

$$\phi' T = \sum_{k\in K'} A_k T A_k^* \;, \qquad \phi'^* X = \sum_{k\in K'} A_k^* X A_k \;. \tag{3.20}$$

The corresponding non-selective operation $\tilde{\phi} = \phi + \phi'$ is thus given by

$$\tilde{\phi} T = \sum_{k\in J} A_k T A_k^* \;, \qquad \tilde{\phi}^* X = \sum_{k\in J} A_k^* X A_k \;. \tag{3.21}$$

Vice versa, given arbitrary index sets J, K, and K' as above, and operators A_k, $k \in J$ satisfying (3.18), then ϕ and ϕ' as defined by (3.19) and (3.20) are operations complementary to each other.-

The operators A_k describing a given operation ϕ according to Theorem 1 (or two complementary operations ϕ and ϕ', according to Theorem 1') are not uniquely determined. To illustrate this, assume ϕ is already given in the form (3.2) with operators A_k, $k \in K$. Choose another index set L, of cardinality not less than that of K, and an "L × K" matrix U with matrix elements u_{lk}, $l \in L$, $k \in K$, which satisfies $U^*U = 1$ (the "K × K" unit matrix) or, more explicitly,

$$\sum_{l\in L} u_{lk}^* u_{lk'} = \delta_{kk'} \quad \text{for all } k,k' \in K \;.$$

(The above assumption about the cardinalities of K and L guarantees the existence of such matrices U.) Replace the operators A_k by new operators

$$B_l = \sum_{k\varepsilon K} u_{lk} A_k \ , \quad l \ \varepsilon \ L \ .$$

Then these operators also satisfy a condition of the form (3.1), because

$$\sum_{l\varepsilon L} B_l^* \ B_l = \sum_{k,k'\varepsilon K, l\varepsilon L} u_{lk}^* \ A_k^* \ u_{lk'} A_{k'}$$

$$= \sum_{k\varepsilon K} A_k^* \ A_k = F \leq 1 \ .$$

Moreover, a similar calculation yields

$$\sum_{l\varepsilon L} B_l \ T \ B_l^* = \sum_{k\varepsilon K} A_k \ T \ A_k^* = \phi T$$

for arbitrary $T \ \varepsilon \ B(H)_1$. Therefore the given operation ϕ may as well be described by the new set of operators B_l, $l \ \varepsilon \ L$, which in general is quite different from the original set of operators A_k, $k \ \varepsilon \ K$.

(To become completely rigorous, the above argument would require additional convergence considerations if either L or both K and L are infinite sets. However, in order to illustrate the non-uniqueness of the representation (3.2), it is sufficient to consider finite index sets only, or - if K is infinite - to choose the above matrix U such that it actually modifies finitely many of the operators A_k only.)

Some conclusions of more physical nature can also be drawn immediately from Theorem 1. First, this theorem strongly supports the point of view that not only projection operations E but, more generally, arbitrary operators $F \ \varepsilon \ L(H)$ represent effects. Indeed, every such operator F corresponds to an operation ϕ in the sense of Theorem 1. (Take, e.g., $K = \{1\}$ and $A_1 = F^{1/2}$.)

Second, Theorem 1 proves that, as mentioned before, the effect F does not uniquely specify the operation ϕ corresponding to it; or, in other words, different effect apparatuses f_1 and f_2 in the same equivalence class F need not perform the same operation. To see this, note that an operation ϕ corresponding to a given effect F may be constructed as follows. Consider an arbitrary decomposition

$$F = \sum_{k\varepsilon K} F_k$$

of F into finitely or infinitely many operators $F_k > 0$. For each $k \varepsilon K$, choose an arbitrary operator U_k with $U_k^* U_k = 1$ (i.e., an isometric operator), and set $A_k = U_k F_k^{1/2}$. Then, indeed,

$$\sum_{k\varepsilon K} A_k^* A_k = \sum_{k\varepsilon K} F_k = F ,$$

which implies that the operators A_k satisfy condition (3.1) and describe an operation ϕ belonging to the effect F. The large degree of arbitrariness in this construction of the operators A_k is obvious. That, moreover, two different sets of operators A_k constructed in this way can really yield two different operations, rather than only two different formal descriptions of the same operation, is easily seen with the help of simple examples. Consider, for instance, the operations ϕ and $\bar{\phi}$ described by

$$K = \{1\}, \quad A_1 = F^{1/2}$$

and

$$\bar{K} = \{1,2\}, \quad \bar{A}_1 = (\tfrac{1}{2}F)^{1/2}, \quad \bar{A}_2 = U(\tfrac{1}{2}F)^{1/2},$$

with an arbitrary isometric operator U, respectively. Both ϕ and $\bar{\phi}$ correspond to the effect F, but $\phi \neq \bar{\phi}$ since ϕ transforms pure states $W = | f \rangle\langle f |$ into pure states $\hat{W} = | \hat{f} \rangle\langle \hat{f} |$, with $\hat{f} = F^{1/2} f / \| F^{1/2} f \|$ – thus being a <u>pure operation</u> in the sense of Ref. [5] – whereas $\bar{\phi}$ transforms the same initial state $W = | f \rangle\langle f |$ into

$$\hat{W} = \frac{1}{2} | \hat{f} \rangle\langle \hat{f} | + \frac{1}{2} | U\hat{f} \rangle\langle U\hat{f} | ,$$

which is not pure unless $U\hat{f} = e^{i\alpha}\hat{f}$.

In particular, operations corresponding to decision effects E are not necessarily described by the "wave packet reduction" formulae (2.8) and (2.9), which represent a very particular case of Eqs. (3.19) to (3.21) with, e.g., $J = \{1,2\}$, $A_1 = E$, $A_2 = E'$, $K = \{1\}$ and $K' = \{2\}$. The very limited validity of Eqs. (2.8) and (2.9) is also an empirical fact, since there are many actual measuring procedures for decision effects E which can not be described by these formulae - even if, as done here, only "non-destructive" measurements are considered.

The "ideal measurement" formulae (2.8) and (2.9) are usually derived with the help of strong additional assumptions. Particularly important among them is the postulate that a repetition of the E measurement in the new state $\hat{W}$ yields the result "yes" with certainty. Measurements fulfilling this postulate are called measurements "of the first kind". In this sense (but slightly more generally), we may consider a selective operation ϕ as constituting a measurement of the first kind of the corresponding effect $F = \phi^*1 \in L(H)$, if

$$tr(F\hat{W}) = tr(F\cdot\phi W)/tr(FW) = tr(\phi^*F\cdot W)/tr(FW) = 1 ,$$

i.e., if

$$tr(\phi^*F\cdot W) = tr(FW)$$

for all states W with non-vanishing transition probabilities tr(FW). But since $tr(FW) = tr(\phi W) = 0$ implies $\phi W = 0$, and thus also $tr(\phi^*F\cdot W) = tr(F\cdot\phi W) = 0$, the last equation must actually hold for all states W. Therefore an operation ϕ constitutes an F measurement of the first kind if (and only if)

$$\phi^*F = F .$$

This condition, clearly, is not valid for arbitrary operations ϕ. (For counterexamples, take $K = \{1\}$, $A_1 = F^{1/2}$, such that $\phi^*F = F^2 \neq F$ unless F is a decision effect E. In the latter case, set $A_1 = UE$ with a suitable unitary U, such that $\phi^*E = EU^*EUE \neq E$.) Since also in practice most measurements are known not to be "of the first kind", there is thus neither a practical nor a theoretical justification for such a postulate.

Not every state transformation of physical interest is an operation. Perhaps the most important counterexample is time reversal, as described by [11]

$$\Theta : W \to \mathbf{T}\, W\, \mathbf{T}^* \tag{3.22}$$

with an antiunitary operator $\mathbf{T}$. (I.e., $\mathbf{T}$ is antilinear,

$$\mathbf{T}(af + bg) = a^*\mathbf{T}f + b^*\mathbf{T}g \ ,$$

and satisfies

$$\mathbf{T}^*\, \mathbf{T} = \mathbf{T}\, \mathbf{T}^* = 1 \ ,$$

with $\mathbf{T}^*$ defined by

$$(f,\mathbf{T}^*g) = (g,\mathbf{T}f)$$

for all f, g ε H.) By (3.22), the mapping Θ has the following properties:

(i) It transforms pure states into pure states.

(ii) It is trace preserving: $tr(\Theta W) = trW$.

(iii) It transforms the set of pure states onto (rather than only into) itself.

If Θ were an operation, it would be a pure one, according to (i), and therefore it could be also represented in the form [12]

$$\Theta : W \to A\, W\, A^* \tag{3.23}$$

with a linear operator A. By (ii) (the "non-selectivity" of Θ), A would be isometric: $A^*A = 1$. Then (iii) would imply that A is even unitary, i.e., also satisfying $AA^* = 1$. (Actually (iii) is also needed to exclude the possibility that Θ, although being pure, might not be of the simple form (3.23) [12].) However, a state mapping of the form (3.22) with antiunitary $\mathbf{T}$ can not be rewritten in the form (3.23) with unitary A [11]. Thus, indeed, time reversal is not an operation. Or, in other words: it is impossible - not only practically, but in principle - to build an apparatus which, when applied to an arbitrary ensemble of microsystems, would turn it into an ensemble in the time reversed state.

The mapping (3.22) of K(H) into itself may easily be extended linearly to $B(H)_1$, and is then given by

$$\Theta : T \to \mathbf{T}\, T^* \,\mathbf{T}^* \, . \tag{3.24}$$

This extended mapping is complex-linear and positive, by construction, and trace preserving. But then it can not be completely positive, since otherwise it would correspond to an operation. Although we could also prove this directly (e.g., by proving that Θ is not even 2-positive), and such proof might even be of some interest from the purely mathematical point of view, we feel no need to present this here.

§4 Composite Systems

With the first representation theorem we have obtained a purely mathematical - although comparatively simple, and thus practically useful - description of arbitrary operations ϕ. In §5 below we shall present a completely different description in the form of our second representation theorem. Although somewhat more complicated mathematically, this theorem can be interpreted in terms of a quantum mechanical model of the measurement process, and is thus perhaps more transparent from the physical point of view. Its formulation and interpretation requires some knowledge of the quantum mechanical description of composite systems. This subject, touched already in §2 and §3, shall thus be discussed now in a more systematic way.

Given two distinct and noninteracting quantum mechanical systems I and II with state spaces H_I and H_{II}, respectively, the state space of the combined system I + II is the product Hilbert space $\underline{H} = H_I \otimes H_{II}$. Density operators $\underline{W} \in K(\underline{H})$, therefore, describe equivalence classes of preparing instruments $\underline{w}$ for pairs of microsystems, each pair consisting of one system I and one system II, and effects $\underline{F} \in L(\underline{H})$ are to be measured by effect apparatuses $\underline{f}$ exposed to, and possibly triggered by, such pairs.

Consider an arbitrary decision effect $E_I \in L(H_I)$ of the isolated system I. Standard quantum mechanics associates with it another decision effect $E_I \otimes 1_{II} \in L(\underline{H})$ of the composite system, with the following physical interpretation. Assume there exists at least one effect aparatus e_I which, when applied to the isolated system I, measures the effect E_I, but which can also be applied to subsystem I of the composite system I + II, without being influenced in that case by the presence of system II. (This property of the apparatus could be tested, e.g., by preparing two ensembles of pairs I + II in the same state $\underline{W}$, removing subsystem II of each pair in one ensemble, and comparing the probabilities for the triggering of the apparatus in these two ensembles.) When applied to pairs I + II, the E_I apparatus e_I has to be considered as an effect apparatus for the composite system; the corresponding effect in $L(\underline{H})$ is assumed to be $\underline{E}_I = E_I \otimes 1_{II}$.

Likewise, an apparatus e_{II} measuring a decision effect E_{II} on system II while being insensitive to system I yields, when applied to the composite system I + II, the decision effect $\underline{E}_{II} = 1_I \otimes E_{II}$ in $L(\underline{H})$.

For arbitrary E_I and E_{II}, the projection operators $\underline{E}_I$ and $\underline{E}_{II}$ on $\underline{H}$ commute. Therefore they describe, again according to standard quantum mechanics, "commensurable" decision effects of the composite system; i.e., these effects can be "measured together", in the following sense. (See also §6.) There exists a suitable apparatus with two different "yes-no pointers", both responding to single systems (pairs) I + II, such that an ensemble of N pairs in a state $\underline{W}$ triggers the first ($\underline{E}_I$) and the second ($\underline{E}_{II}$) pointer approximately $\mathrm{tr}(\underline{E}_I\underline{W})N$ and $\mathrm{tr}(\underline{E}_{II}\underline{W})N$ times, respectively. This apparatus can also be used to perform the yes-no measurement "$\underline{E}_I$ and $\underline{E}_{II}$", defined - as in ordinary logic (Boolean algebra) - to give the result "yes" if both pointers are triggered together by the same single pair I + II. Quantum mechanics associates with this correlation measurement "$\underline{E}_I$ and $\underline{E}_{II}$" the product

$$\underline{E}_I \cdot \underline{E}_{II} = (E_I \otimes 1_{II})(1_I \otimes E_{II}) = E_I \otimes E_{II} \tag{4.1}$$

of the operators $\underline{E}_I$ and $\underline{E}_{II}$, i.e., another decision effect $\underline{E}_{I,II} = E_I \otimes E_{II}$ of the composite system.

In the particular case considered, an apparatus for the joint measurement of $\underline{E}_I$ and $\underline{E}_{II}$ can be realized as follows. Assume there exists an apparatus e_I measuring, as above, the effects E_I on isolated systems I and $\underline{E}_I$ on composite systems I + II, and which is not only insensitive to, but also does not influence at all the subsystem II in the latter case. (In view of the absence of interactions between the subsystems I and II, the existence of such an apparatus is at least not implausible. The operation performed by such an apparatus is discussed below.) One then expects that a combination of such an E_I apparatus e_I with an analogous E_{II} apparatus e_{II} for subsystem II measures $\underline{E}_I$ and $\underline{E}_{II}$ together on the composite system. This is indeed confirmed by the detailed theory presented below.

A particular type of state $\underline{W}$ of the composite system can be prepared as follows. Take a preparing instrument w_I for system I and another one, w_{II}, for system II. Prepare N systems with each apparatus, and combine them into N pairs I + II. The state $\underline{W}$ of such an ensemble of N pairs is called an uncorrelated state.

In view of this preparing procedure for $\underline{W}$ and the above-described explicit construction of an apparatus for the combined measurement of $\underline{E}_I$ and $\underline{E}_{II}$, it is intuitively obvious that the decision effect "$\underline{E}_I$ and $\underline{E}_{II}$" (Eq. (4.1)) occurs in an uncorrelated state $\underline{W}$ with the probability

$$\mathrm{tr}((E_I \otimes E_{II})\underline{W}) = \mathrm{tr}(E_I W_I)\cdot \mathrm{tr}(E_{II} W_{II}) \; ; \qquad (4.2)$$

i.e., the probability for the joint occurrence of the "subsystem" effects $\underline{E}_I$ and $\underline{E}_{II}$ is the product of the probabilities of the effects E_I and E_{II} in the subsystem states W_I and W_{II} prepared by the instruments w_I and w_{II}, respectively. The two last-mentioned probabilities coincide with the probabilities of the effects $\underline{E}_I$ and $\underline{E}_{II}$ in the composite system state $\underline{W}$ since, with either E_{II} or E_I replaced by the unit operator, (4.2) implies

$$\mathrm{tr}(\underline{E}_I\underline{W}) = \mathrm{tr}(E_I W_I) \; , \quad \mathrm{tr}(\underline{E}_{II}\underline{W}) = \mathrm{tr}(E_{II} W_{II}) \; . \qquad (4.3)$$

Like (4.2), these relations also have a very simple and intuitively obvious physical interpretation.

Eq. (4.2) implies that the density operator $\underline{W}$ of an uncorrelated state is the tensor product

$$\underline{W} = W_I \otimes W_{II} \tag{4.4}$$

of the density operators W_I and W_{II} describing the subsystem states of which $\underline{W}$ is "composed". This is because the right hand side of (4.2) is identical to, and can thus be replaced by, $\mathrm{tr}\big((E_I \otimes E_{II})(W_I \otimes W_{II})\big)$. For one-dimensional projection operators $E_I = |f_I \rangle\langle f_I|$ and $E_{II} = |f_{II} \rangle\langle f_{II}|$, for which

$$E_I \otimes E_{II} = |f_I \otimes f_{II} \rangle\langle f_I \otimes f_{II}| ,$$

(4.2) can then be rewritten, with

$$\underline{A} = \underline{W} - W_I \otimes W_{II} ,$$

in the form

$$\big((f_I \otimes f_{II}), \underline{A}(f_I \otimes f_{II})\big) = 0 . \tag{4.5}$$

This is true for arbitrary unit vectors $f_I \varepsilon H_I$ and $f_{II} \varepsilon H_{II}$. The restriction to unit vectors is easily removed by multiplying (4.5) with suitable scale factors. Applying now the polarization identity (1.13) for $A = \underline{A}$, $f = f_I \otimes f_{II}$, and $g = g_I \otimes f_{II}$, with $g_I \varepsilon H_I$ arbitrary, we obtain from (4.5)

$$\big((f_I \otimes f_{II}), \underline{A}(g_I \otimes f_{II})\big) = 0 .$$

A similar identity,

$$4\big((f_I \otimes f_{II}), \underline{A}(g_I \otimes g_{II})\big) = \big((f_I \otimes (f_{II}+ g_{II})), \underline{A}(g_I \otimes (f_{II}+ g_{II}))\big)$$
$$- \big((f_I \otimes (f_{II}- g_{II})), \underline{A}(g_I \otimes (f_{II}- g_{II}))\big)$$
$$+ i\big((f_I \otimes (f_{II}-ig_{II})), \underline{A}(g_I \otimes (f_{II}-ig_{II}))\big)$$
$$- i\big((f_I \otimes (f_{II}+ig_{II})), \underline{A}(g_I \otimes (f_{II}+ig_{II}))\big) ,$$

then leads to

$$\big((f_I \otimes f_{II}), \underline{A}(g_I \otimes g_{II})\big) = 0 ,$$

valid for all $f_I, g_I \in H_I$ and $f_{II}, g_{II} \in H_{II}$. Since finite linear combinations of product vectors are dense in $\underline{H}$, this finally implies $\underline{A} = 0$, i.e., Eq. (4.4).

Consider now, more generally, an effect $F_I \in L(H_I)$ of the isolated system I which need not be a decision effect. Assume, as above for the particular case $F_I = E_I$, that there exists at least one effect apparatus f_I in the equivalence class F_I which can also be applied to composite systems I + II, without affecting nor being affected by system II in this case. If applied in this way, the apparatus f_I measures a certain effect $\underline{F}_I \in L(\underline{H})$ of the composite system. By this operational definition, the effect $\underline{F}_I$ must occur in an uncorrelated state (4.4) with a probability

$$\mathrm{tr}(\underline{F}_I(W_I \otimes W_{II})) = \mathrm{tr}(F_I W_I) , \qquad (4.6)$$

coinciding with the probability for the effect F_I in the state W_I of subsystem I. Likewise, suitable effect apparatuses f_{II} corresponding to effects F_{II} of the isolated subsystem II define effects $\underline{F}_{II} \in L(\underline{H})$, with

$$\mathrm{tr}(\underline{F}_{II}(W_I \otimes W_{II})) = \mathrm{tr}(F_{II} W_{II}) , \qquad (4.7)$$

when applied to the composite system. Finally, the combination of two such apparatuses f_I and f_{II} is again expected to measure $\underline{F}_I$ and $\underline{F}_{II}$ together. (Thus $\underline{F}_I$ and $\underline{F}_{II}$ are "coexistent" effects, as discussed in detail in §6.) The same combined apparatus can then also be used to measure the effect $\underline{F}_{I,II}$ = "$\underline{F}_I$ and $\underline{F}_{II}$", defined (as above) to occur if and only if both f_I and f_{II} are triggered together by the same single pair I + II. Again the probability for $\underline{F}_{I,II}$, i.e., for the joint occurrence of $\underline{F}_I$ and $\underline{F}_{II}$, must factorize in the form

$$\mathrm{tr}(\underline{F}_{I,II}(W_I \otimes W_{II})) = \mathrm{tr}(F_I W_I)\cdot\mathrm{tr}(F_{II} W_{II}) \tag{4.8}$$

for uncorrelated states. Inserting, in particular, pure states $W_I = |f_I\rangle\langle f_I|$ and $W_{II} = |f_{II}\rangle\langle f_{II}|$, Eq. (4.8) now implies

$$\underline{F}_{I,II} = F_I \otimes F_{II} , \tag{4.9}$$

while Eqs. (4.6) and (4.7) imply

$$\underline{F}_I = F_I \otimes 1_{II} , \quad \underline{F}_{II} = 1_I \otimes F_{II} . \tag{4.10}$$

(Compare the derivation of (4.4) from (4.2), and note that (4.6) and (4.7) are just particular cases of (4.8). See also Eq. (3.8) of §3.) We have thus shown that relations which are already known for decision effects (e.g., (4.1)) can be generalized immediately, in the form of Eqs. (4.9) and (4.10), to arbitrary effects F_I and F_{II}. Effects of the form (4.10) and (4.9) will be called subsystem and correlation effects, respectively, in the following.

For an arbitrary state $\underline{W} \in K(\underline{H})$ of the composite system there exists, as is well known, a unique density operator $\mathrm{Tr}_{II}\underline{W} \in K(H_I)$ for subsystem I, called the reduction of the state $\underline{W}$ to that subsystem, such that

$$\mathrm{tr}((F_I \otimes 1_{II})\underline{W}) = \mathrm{tr}(F_I \cdot \mathrm{Tr}_{II}\underline{W}) \tag{4.11}$$

for all $F_I \in L(H_I)$. Thus $Tr_{II}\underline{W}$ describes the statistics of arbitrary subsystem I effects in the given state $\underline{W}$; or, in other words, $Tr_{II}\underline{W}$ describes the state of the N subsystems I in an ensemble of N pairs I + II in the state $\underline{W}$.

More generally, there exists for any $\underline{T} \in B(\underline{H})_1$ a unique operator $T_I = Tr_{II}\underline{T} \in B(H_I)_1$, which is defined implicitly by requiring

$$tr(X_I \cdot Tr_{II}\underline{T}) = tr((X_I \otimes 1_{II})\underline{T}) \tag{4.12}$$

for all $X_I \in B(H_I)$. Namely, for fixed $\underline{T}$, the right hand side of (4.12) defines a linear functional $\tau(X_I)$ on $B(H_I)$, which is norm continuous since

$$|\tau(X_I)| \leq \|X_I \otimes 1_{II}\| \, \|\underline{T}\|_1 = \|\underline{T}\|_1 \, \|X_I\| ,$$

and therefore is of the form $tr(X_I T_I)$ with a unique $T_I \in B(H_I)_1$. Eq. (4.12) thus defines a mapping $Tr_{II} : B(\underline{H})_1 \to B(H_I)_1$ which, obviously, is linear. It is also positive since, by (4.12), $\underline{T} \geq 0$ implies $tr(X_I \cdot Tr_{II}\underline{T}) \geq 0$ for all $X_I \geq 0$, and thus $Tr_{II}\underline{T} \geq 0$. This also implies $Tr_{II}(\underline{T}^*) = (Tr_{II}\underline{T})^*$; i.e., Tr_{II} is a real mapping, and (4.12) with $X_I = 1_I$ yields $tr(Tr_{II}\underline{T}) = tr\underline{T}$; i.e., Tr_{II} is trace preserving. Therefore Tr_{II} maps states $\underline{W} \in K(\underline{H})$ into states $Tr_{II}\underline{W} \in K(H_I)$, and (4.11) follows as a special case of (4.12). For $\underline{T} = T_I \otimes T_{II}$, (4.12) implies

$$tr(X_I \cdot Tr_{II}\underline{T}) = tr(X_I T_I) \cdot trT_{II} ,$$

and thus

$$Tr_{II}(T_I \otimes T_{II}) = trT_{II} \cdot T_I \tag{4.13}$$

for arbitrary $T_I \in B(H_I)_1$, $T_{II} \in B(H_{II})_1$. This also implies that Tr_{II} maps $B(\underline{H})_1$ onto $B(H_I)_1$, and $K(\underline{H})$ onto $K(H_I)$.

A more explicit equation for $Tr_{II}\underline{T}$ follows from (4.12) by inserting $X_I = |f_I\rangle\langle f_I|$ with an arbitrary unit vector $f_I \in H_I$. Evaluating the trace on the left hand side with a particular orthonormal basis $\{f_I^i\}$ in H_I, for which $f_I^1 = f_I$, and the trace on the right hand side with the basis $\{f_I^i \otimes g_{II}^k\}$ in $H_I \otimes H_{II}$, with $\{f_I^i\}$ as before and an arbitrary basis $\{g_{II}^k\}$ in H_{II}, we then get from (4.12)

$$(f_I, \mathrm{Tr}_{II}\underline{T}\, f_I) = \sum_k ((f_I \otimes g_{II}^k), \underline{T}(f_I \otimes g_{II}^k)) .$$

By multiplying with scale factors and applying the polarization identity (1.13), we obtain from this the well-known formula

$$(f_I, \mathrm{Tr}_{II}\underline{T}\, g_I) = \sum_k ((f_I \otimes g_{II}^k), \underline{T}(g_I \otimes g_{II}^k)) , \qquad (4.14)$$

valid for arbitrary $f_I, g_I \in H_I$. Thus $\mathrm{Tr}_{II}\underline{T}$ results from $\underline{T}$ by performing a "partial trace" with respect to H_{II}; hence our notation.

By interchanging the rôles of H_I and H_{II}, we obtain another mapping $\mathrm{Tr}_I : B(\underline{H})_1 \to B(H_{II})_1$, the partial trace with respect to H_I, with analogous properties and a similar physical interpretation.

An effect apparatus f_I, performing a selective operation ϕ_I by measuring the effect $F_I = \phi_I^* 1_I$ on subsystem I and not interacting with subsystem II, can also be used to perform a selective operation $\underline{\phi}_I$ on the composite system I + II. In this case, pairs I + II are selected according to the occurrence of the subsystem I effect $\underline{F}_I = F_I \otimes 1_{II}$, as measured by f_I when applied to such pairs. If ϕ_I is represented in the form (cf. Theorem 1)

$$\phi_I W_I = \sum_{k\in K} A_{Ik}\, W_I\, A_{Ik}^*$$

with suitable operators A_{Ik} $(k \in K)$ on H_I, the operation $\underline{\phi}_I$ is given by

$$\underline{\phi}_I \underline{W} = \sum_{k\in K} (A_{Ik} \otimes 1_{II})\, \underline{W}\, (A_{Ik} \otimes 1_{II})^* . \qquad (4.15)$$

Then, indeed, the effect corresponding to $\underline{\phi}_I$ is

$$\sum_{k\varepsilon K}(A_{Ik}\otimes 1_{II})^*(A_{Ik}\otimes 1_{II}) = (\sum_{k\varepsilon K} A^*_{Ik}A_{Ik})\otimes 1_{II} = F_I \otimes 1_{II} = \underline{F}_I \ ,$$

in accordance with (4.10), whereas for uncorrelated states we have

$$\underline{\phi}_I(W_I \otimes W_{II}) = (\sum_{k\varepsilon K} A_{Ik}W_IA^*_{Ik})\otimes W_{II} = \phi_I W_I \otimes W_{II} \ , \tag{4.16}$$

as expected form the operational definition of $\underline{\phi}_I$. (Compare also Eqs. (3.6) to (3.8) of §3.)

One can also show that, conversely, every operation $\underline{\phi}_I$ which acts on subsystem I of the composite system I + II only, and therefore transforms uncorrelated states in accordance with Eq. (4.16), indeed must be of the form (4.15). The proof is rather lengthy, however, and is therefore omitted here.

With $\underline{\phi}_I$ as given by (4.15) and arbitrary $X_I \varepsilon B(H_I)$ and $X_{II} \varepsilon B(H_{II})$, we have

$$\underline{\phi}^*_I(X_I \otimes X_{II}) = (\sum_{k\varepsilon K} A^*_{Ik}X_IA_{Ik})\otimes X_{II} = \phi^*_I X_I \otimes X_{II} \ .$$

From this and Eq. (4.12) we get, for arbitrary $\underline{W} \varepsilon K(\underline{H})$,

$$\begin{aligned} tr(X_I\cdot Tr_{II}(\underline{\phi}_I\underline{W})) &= tr((X_I \otimes 1_{II})\cdot\underline{\phi}_I\underline{W}) \\ &= tr(\underline{\phi}^*_I(X_I \otimes 1_{II})\cdot\underline{W}) = tr((\phi^*_I X_I \otimes 1_{II})\cdot\underline{W}) \\ &= tr(\phi^*_I X_I\cdot Tr_{II}\underline{W}) = tr(X_I\cdot\phi_I(Tr_{II}\underline{W})) \ . \end{aligned}$$

Since X_I is arbitrary, we finally obtain

$$Tr_{II}(\underline{\phi}_I\underline{W}) = \phi_I(Tr_{II}\underline{W}) \ . \tag{4.17}$$

This equation means, physically, that the final state of subsystem I after the operation $\underline{\phi}_I$ - as given, up to a normalization factor, by $Tr_{II}(\underline{\phi}_I\underline{W})$ - may also be obtained by applying the operation ϕ_I to the initial state $\mathrm{Tr}_{II}\underline{W}$ of subsystem I. By the operational definition of $\underline{\phi}_I$, this must indeed be true for arbitrary initial states $\underline{W}$ of the composite system. (For uncorrelated states, this property of $\underline{\phi}_I$ is already expressed by (4.16).)

The measurement of the correlation effect $\underline{F}_{I,II}$ = "$\underline{F}_I$ and $\underline{F}_{II}$" in an ensemble of N >> 1 pairs I + II in a given state $\underline{W}$ by means of a combined aparatus, consisting of f_I and a similar apparatus f_{II} acting on subsystem II, may now also be described as follows. The apparatus f_I is triggered by

$$N_I = N\, \mathrm{tr}(\underline{F}_I \underline{W}) = N\, \mathrm{tr}(\underline{\phi}_I \underline{W})$$

pairs, which form an ensemble in the new state

$$\hat{\underline{W}} = \underline{\phi}_I \underline{W} / \mathrm{tr}(\underline{\phi}_I \underline{W}) \; .$$

In this ensemble, the subsystem II effect $\underline{F}_{II} = 1_I \otimes F_{II}$ measured by f_{II} then occurs

$$N_{I,II} = N_I\, \mathrm{tr}(\underline{F}_{II} \hat{\underline{W}}) = N\, \mathrm{tr}(\underline{F}_{II} \cdot \underline{\phi}_I \underline{W}) = N\, \mathrm{tr}(\underline{\phi}_I^* \underline{F}_{II} \cdot \underline{W})$$

times. Thus f_I and f_{II} are triggered together, in an arbitrarily given state $\underline{W}$, with the probability $N_{I,II}/N = \mathrm{tr}(\underline{F}_{I,II} \cdot \underline{W})$, where

$$\begin{aligned} \underline{F}_{I,II} = \underline{\phi}_I^* \underline{F}_{II} &= \sum_{k\in K} (A_{Ik} \otimes 1_{II})^* \, (1_I \otimes F_{II}) \, (A_{Ik} \otimes 1_{II}) \\ &= (\sum_{k\in K} A_{Ik}^* \, A_{Ik}) \otimes F_{II} = F_I \otimes F_{II} \; , \end{aligned}$$

in accordance with (4.9).

This interpretation of the $\underline{F}_{I,II}$ measurement is appropriate if the parts f_I and f_{II} of the combined apparatus are applied to, and possibly triggered by, each single pair I + II in the order considered (i.e., first f_I, then f_{II}). If, vice versa, the apparatus f_{II} is applied before f_I, a completely analogous consideration with f_I and f_{II} interchanged again leads to (4.9). Our previous derivation of

Eq. (4.9) from the requirement (4.8) for the joint triggering of f_I and f_{II} in uncorrelated states is more general, however, and in particular also independent of the temporal order of the two measurements performed by f_I and f_{II}.

According to (4.16), the operation $\underline{\phi}_I$ does not change the state W_{II} of subsystem II when applied to an uncorrelated state $W_I \otimes W_{II}$. This is not true for more general states $\underline{W}$; in fact, one may easily construct examples for which the final state $\mathrm{Tr}_I\hat{\underline{W}}$ of subsystem II, with $\hat{\underline{W}} = \underline{\phi}_I\underline{W}/\mathrm{tr}(\underline{\phi}_I\underline{W})$, is different from its initial state $\mathrm{Tr}_I\underline{W}$. This is by no means surprising. If in the state $\underline{W}$ there are correlations between the subsystems I and II, then a selection according to a subsystem I effect $\underline{F}_I$ may indeed change the state of subsystem II. Such state changes are thus not due to an interaction between the apparatus f_I and subsystem II, but rather result from the interplay of the correlations and the applied selection procedure. Therefore they must also be absent if no selection is made, i.e., if instead of $\underline{\phi}_I$ the corresponding non-selective operation $\tilde{\underline{\phi}}_I$ is considered.

According to Theorem 1' and an obvious generalization of Eq. (4.15), the non-selective operation $\tilde{\underline{\phi}}_I$ performed by the apparatus f_I on the composite system transforms $\underline{W}$ into the new state

$$\tilde{\underline{\phi}}_I\underline{W} = \sum_{k\varepsilon J} (A_{Ik} \otimes 1_{II})\, \underline{W}\, (A_{Ik} \otimes 1_{II})^* \ .$$

Here J is an index set of which the set K in (4.15) is a subset, and the operators A_{Ik} $(k \varepsilon J)$ on H_I satisfy

$$\sum_{k\varepsilon J} A^*_{Ik}\, A_{Ik} = 1_I \ .$$

For arbitrary subsystem II effects $\underline{F}_{II} = 1_I \otimes F_{II}$ we then get

$$\tilde{\underline{\phi}}^*_I(1 \otimes F_{II}) = \sum_{k\varepsilon J} (A_{Ik} \otimes 1_{II})^* (1_I \otimes F_{II}) (A_{Ik} \otimes 1_{II})$$

$$= (\sum_{k\varepsilon J} A^*_{Ik}\, A_{Ik}) \otimes F_{II} = 1_I \otimes F_{II} \ ,$$

and thus

$$\mathrm{tr}((1_I \otimes F_{II})\cdot\tilde{\Phi}_I\underline{W}) = \mathrm{tr}(\tilde{\Phi}_I^*(1_I \otimes F_{II})\cdot\underline{W}) = \mathrm{tr}((1_I \otimes F_{II})\cdot\underline{W}) \ .$$

This means (compare Eq. (4.11))

$$\mathrm{Tr}_I(\tilde{\Phi}_I\underline{W}) = \mathrm{Tr}_I\underline{W} \ ;$$

i.e., the reduction of the composite system state to subsystem II is indeed unchanged by the operation $\tilde{\Phi}_I$.

The preceding discussion should be sufficient to demonstrate the internal consistency of the formal description of composite systems without interaction. For later use, however, we have to generalize the theory to composite systems I + II whose subsystems I and II interact with each other.

The above operational definition of subsystem and correlation effects can not be generalized immediately to the case of interacting subsystems, since then, obviously, one can not apply effect apparatuses to one subsystem without perturbing the other one. Likewise, the possibility of preparing uncorrelated states by independent preparations of the subsystems becomes questionable. In order to circumvent difficulties of this kind, we restrict our attention to so called "scattering systems", which are similar enough to noninteracting ones to be treated explicitly in a simple way, and which on the other hand are just the kind of systems encountered later on in §5.

A quantum mechanical system is called here a binary scattering system, if it behaves in both the distant past and future as a composite system of two noninteracting subsystems I and II, which can as well be prepared and studied independently of each other. If there were no interaction at all, the state space would be $H_I \otimes H_{II}$, and there would exist particular subsystem and correlation effects (Eqs. (4.10) and (4.9)) with the physical interpretation discussed above. Given an arbitrary state $\underline{W}$ of the composite system, it would suffice to measure the probabilities $\mathrm{tr}(\underline{F}_{I,II}\underline{W})$ for sufficiently many correlation effects (4.9) in order to determine $\underline{W}$ completely. (Take, e.g., $F_I = |f_I \rangle\langle f_I|$ and $F_{II} = |f_{II} \rangle\langle f_{II}|$, so that $tr(\underline{F}_{I,II}\underline{W}) = ((f_I \otimes f_{II}), \underline{W}(f_I \otimes f_{II}))$, and recall the derivation of Eq. (4.4).)

The state space $\underline{H}$ of a binary scattering system may still be identified with $H_I \otimes H_{II}$, and there also exist particular "correlation" effects of the form (4.9) - including, for $F_{II} = 1_{II}$ or $F_I = 1_I$, the "subsystem" effects (4.10) - with the following physical meaning. Given an ensemble of such systems in some state $\underline{W}$, there exists - by the definition of a binary scattering system - a time T_- before which each system of the ensemble behaves like a pair of two noninteracting subsystems I and II. This time T_- will depend, in general, on the state $\underline{W}$ of the ensemble. Effect apparatuses constructed for one subsystem and not interacting with the other one can thus also be applied, at times $t < T_-$, to all scattering systems in the given ensemble. (In most cases the subsystems I and II are spatially separated form each other for $t < T_-$, which clearly facilitates separate measurements on them.) Such separate measurements are now described by operators of the form (4.10), while the joint occurrence of two such effects when measured together corresponds to an effect of the form (4.9). Sufficiently many measurements of this type again uniquely determine the statistical operator $\underline{W}$ on $\underline{H} = H_I \otimes H_{II}$.

Certain difficulties seem to arise here, however. First, the state $\underline{W}$ (i.e., the corresponding ensemble of scattering systems) might have been prepared at a time later than T_-, so that it describes an ensemble of interacting systems from the very beginning, and therefore no measurements at all could be performed before the onset of the interaction. Second, even if $\underline{W}$ was prepared before the time T_-, it might seem impossible to measure correlation effects $F_I \otimes F_{II}$ with sufficiently many F_I and F_{II} in the remaining limited time interval between the preparation and the onset of the interaction at time T_-.

In conventional quantum mechanics one more or less explicitly assumes, however, that a given effect F can be measured "in principle" by many different apparatuses f including, in particular, one apparatus "operating" in an arbitrarily small time interval around an arbitrarily given time t. Consider, e.g., a free particle, and take for F a characteristic function $\chi_V(\mathbf{X}_o)$ of its position operator $\mathbf{X}_o$ at time t = 0. Here V is a given spatial volume, and its characteristic function χ_V is defined by

$$\chi_V(\mathbf{x}) = \begin{cases} 1 & \text{for } \mathbf{x} \in V \,, \\ 0 & \text{for } \mathbf{x} \notin V \,. \end{cases}$$

Then $\chi_V(\mathbf{X}_o)$ is a projection operator, thus describing a decision effect E. The simplest apparatus measuring E (at least approximately) is a counter occupying the volume V and "switched on" during a small time interval around t = 0. The solutions of Heisenberg's equations of motion for position and momentum of a free particle,

$$\mathbf{X}_t = \mathbf{X}_o + \frac{1}{m}\mathbf{P}_o t \,, \quad \mathbf{P}_t = \mathbf{P}_o \,,$$

imply, however, that E may also be represented as

$$E = \chi_V(\mathbf{X}_t - \frac{1}{m}\mathbf{P}_t t)$$

in terms of position $\mathbf{X}_t$ and momentum $\mathbf{P}_t$ at any other time t. An apparatus measuring this particular function of $\mathbf{X}_t$ and $\mathbf{P}_t$, and thus "operating" at (or at least around) the time t, might thus as well be used to measure E. Although this indicates that "in principle" E should also be measureable at time t, the actual construction of the corresponding apparatus can not be deduced theoretically, and might in fact be quite difficult in practice.

An assumption of this type, if made for the subsystems I and II, removes the second difficulty mentioned above, since then arbitrary subsystem and correlation effects can indeed be measured in any given time interval before T_-. The first difficulty is removed by an analogous assumption for preparing instruments: Given an arbitrary state $\underline{W}$, there shall exist instruments $\underline{w}$ preparing this state and operating at arbitrarily prescribed (in particular, at arbitrarily early) times.

In order to prevent misunderstandings, we want to stress here that we do not assume preparations or measurements to be "instantaneous" in any sense. Such an assumption would, in fact, be quite unrealistic in view of actual preparing and measuring processes. It is also unnecessary for the theory. Nevertheless, such processes always have a finite duration and a definite time of application; e.g., the time when one "switches on" an accelerator, "sensibilizes" a counter, or "exposes" a photographic plate. The time of application is to be included in the specification of the preparing or measuring instrument; in other words, two instruments of the same construction but applied at different times are treated here as different. (This implies that we use the Heisenberg picture here; see below.) We should also mention that assumptions of the type described above enter the theory, more or less explicitly, at other places also. Without such assumptions it would already be difficult to understand how arbitrary effects F could be measured in arbitrary states W. But even the very defintion of effects F and states W in terms of equivalence classes of instruments requires more care, if one takes into account that certain preparing and measuring instruments cannot actually be combined in a single experiment since, e.g., they occupy the same region in space, or the effect apparatus operates earlier than the preparing instrument, etc. To present a theory which properly takes into account all these subtleties goes, however, far beyond the scope of the present discussion.

When analyzed in terms of the correlation effects (4.9), some particular states $\underline{W}$ of the scattering system will turn out to be uncorrelated in the sense of Eq. (4.8), and therefore to be described by $\underline{W} = W_I \otimes W_{II}$. Any such state may also be prepared in the way described above for noninteracting subsystems, i.e., by means of a preparing instrument w_I for the state W_I of subsystem I and another one, w_{II}, for W_{II}. This procedure requires the use - and thus presupposes the existence - of particular preparing instruments w_I and w_{II}, both operating at times before the onset of interactions in the given state $\underline{W}$.

Up to now we have considered the "incoming" subsystem and correlation effects (4.10) and (4.9) only. But since a binary scattering system in a given state $\underline{W}$ behaves like a pair I + II of two noninteracting subsystems in the distant future as well, later than some time T_+ say, on can also operationally define analogous "outgoing" subsystem and correlation effects, now to be measured at times later than T_+. (Like T_-, T_+ also depends on the state $\underline{W}$. Both T_+ and T_- are ambiguous to some extent, and may not even exist in a literal sense if the interaction does not vanish exactly but only becomes "nondetectably weak" for "sufficiently" early and late times. Although for most states T_- and T_+ will satisfy $T_- < T_+$, thus including a more or less well-defined time interval of interaction, there may also exist states without any detectable interaction at all, for which one could set $T_- = +\infty$ and $T_+ = -\infty$.)

Being thus defined in a different way, the outgoing effects describing separate or joint measurements on the subsystems at sufficiently late times do not coincide with the corresponding incoming effects (4.10) or (4.9). A subsystem I apparatus f_I, for instance, in general will be triggered in a given state $\underline{W}$ with different probabilities before and after the interaction; these two measurements therefore can not be described by the same operator $\underline{F}$. However, since nevertheless the physical situations before and after the interaction are analogous, one can also "identify" the state space $\underline{H}$ of the scattering system with the tensor product $H_I \otimes H_{II}$ in such a way, that the outgoing correlation and subsystem effects are described by operators of product form, as in Eqs. (4.9) and (4.10). This "identification" - mathematically, an isomorphism $\mathbf{T}_+$ of $\underline{H}$ onto $H_I \otimes H_{II}$ - can not be the same as the one used previously, which led to the represenation (4.9) and (4.10) for the incoming effects. Actually, we have not introduced explicitly the "in" analog $\mathbf{T}_-$ of $\mathbf{T}_+$, but simply assumed $\underline{H} \equiv H_I \otimes H_{II}$ (so that $\mathbf{T}_-$ became the identity map), in order to simplify our notation. Therefore, $\mathbf{T}_+$ becomes an isomorphism of $\underline{H} = H_I \otimes H_{II}$ onto the same space $H_I \otimes H_{II}$; i.e., a unitary operator $\underline{S}$ on $H_I \otimes H_{II}$. By definition of $\mathbf{T}_+$, the outgoing correlation effect defined in analogy to

(4.9) is represented on $\mathbf{T}_+\underline{H} = H_I \otimes H_{II}$ by the product operator $F_I \otimes F_{II}$. If transformed back to $\underline{H}$, this operator acts like

$$\mathbf{T}_+^{-1}(F_I \otimes F_{II})\mathbf{T}_+ = \underline{S}^*(F_I \otimes F_{II})\underline{S} \ .$$

Outgoing correlation effects are therefore represented in our state space $\underline{H}$ by operators of the form

$$\underline{F}_{I,II}^{out} = \underline{S}^*(F_I \otimes F_{II})\underline{S} \ , \tag{4.18}$$

which differ from the corresponding incoming effects $\underline{F}_{I,II}^{in} \equiv \underline{F}_{I,II}$ as given by (4.9) by a fixed unitary transformation. Likewise, the outgoing subsystem effects corresponding to (4.10) are

$$\underline{F}_{I}^{out} = \underline{S}^*(F_I \otimes 1_{II})\underline{S} \ , \quad \underline{F}_{II}^{out} = \underline{S}^*(1_I \otimes F_{II})\underline{S} \ . \tag{4.19}$$

The operator $\underline{S}$, representing the overall effect of the interaction, is nothing else than the usual scattering operator or S matrix. Its knowledge permits the prediction of the probabilities

$$\mathrm{tr}(\underline{F}_{I,II}^{out}\ \underline{W}) = \mathrm{tr}(\underline{S}^*(F_I \otimes F_{II})\underline{S}\ \underline{W}) \tag{4.20}$$

for arbitrary outgoing correlation (and subsystem) effects, provided the state $\underline{W}$ is also known - e.g., from the measurement of sufficiently many incoming correlation effects, or from the fact that $\underline{W}$ has been prepared as a known uncorrelated state in the distant past. (Predictions of this kind are usually cast in the form of scattering cross sections.)

The quantum mechanical description of a binary scattering system given here is rather incomplete, of course. We have discussed very particular types of yes-no measurements only, which moreover become ill-defined - both physically and mathematically - when applied to the system during the time interval of interaction between T_- and T_+. Nevertheless, this type of description is sometimes quite adequate and sufficient - e.g., in scattering theory, or here in §5.

Quantum mechanics in the form presented here uses the so called Heisenberg picture. Indeed, as is characteristic of that picture, a given state is described here by a fixed statistical operator W. Moreover, the complete specification of an effect apparatus includes the time of application; yes-no measurements performed with the same instrument at different times are thus considered as different. (The corresponding operators F are usually related to each other by unitary time translation operators.) The operator $\underline{S}$ introduced above is thus the Heisenberg picture S matrix.

If one discusses only measurements on the noninteracting incoming and outgoing subsystems of scattering systems, as we are doing here, then the Dirac or interaction pictures is also quite appropriate. In this picture, one rewrites the probabilties (4.20) for outgoing correlation effects in the form

$$\mathrm{tr}((F_I \otimes F_{II})\ \underline{S}\ \underline{W}\ \underline{S}^*) = \mathrm{tr}(\underline{F}_{I,II}\ \underline{W}^{out})$$

with

$$\underline{W}^{out} = \underline{S}\ \underline{W}\ \underline{S}^* \ . \tag{4.21}$$

Accordingly, one describes both incoming and outgoing correlation effects by the same operators $\underline{F}_{I,II}$ given by (4.9), whereas a given state (i.e., a given preparation procedure) is described by two different statistical operators: by

$$\underline{W}^{in} \equiv \underline{W} \tag{4.22}$$

if used to calculate probabilities for incoming effects in the form $\mathrm{tr}(\underline{F}_{I,II}\underline{W}^{in})$, and by $\underline{W}^{out}$, as given by (4.21), which yields the probabilities $\mathrm{tr}(\underline{F}_{I,II}\underline{W}^{out})$ for outgoing effects. In a sense, one thus considers an incomig and the corresponding outgoing effect as "the same", and explains their (in general) different probabilities by a state change $\underline{W}^{in} \to \underline{W}^{out}$ due to the interaction between subsystems I and II in the time interval separating "incoming" and "outgoing" measurements.

This description has the formal advantage that subsystem and correlation effects have the same simple product form both before and after the interaction. (They still do not make sense operationally, however, in the interaction interval.) But this is achieved at the price of spoiling the unique assignment of a statistical operator $\underline{W}$ to a given preparation procedure - i.e., at the cost of conceptual simplicity. From this point of view, the Schrödinger picture is even less appealing, and shall thus not be described here at all.

§5 The Second Representation Theorem

We are now ready to derive, as announced previously, a new mathematical representation of arbitrary operations. This representation is most easily obtained from a quantum mechanical model of a yes-no measurement and the corresponding operations.

Assume for this purpose that the effect apparatus f used in such a measurement can also be described quantum mechanically. Its state space is denoted by H_a, with statistical and effect operators marked by the same subscript a. The microsystem, to which the apparatus is applied, is described in a state space denoted by H, as before. A yes-no measurement performed by the apparatus f on the microsytem in a state $W \in K(H)$ may then be described as follows. Before the measurement, the microsystem and the apparatus do not interact with each other, and are independently prepared in states W and W_a, respectively. Here W_a is the state in which the apparatus f is "ready for measurement", and is thus determined once and for all by the construction of the apparatus f, whereas the state W of the microsystem may be chosen arbitrarily. Then the microsystem and the apparatus interact with each other during a certain time interval (which in general will depend on W unless this interaction is "switched on and off" artificially), and separate again from each other afterwards if - as assumed here - the apparatus f does not "absorb" or "destroy" the microsystem. The occurrence or non-occurrence of the effect measured by the apparatus f is to be determined by a certain yes-no measurement on f (e.g., by "looking at a pointer") after the interaction with the microsystem. This measurement is described formally by some effect operator $F_a \in L(H_a)$, which is again fixed by the construction of the apparatus f.

In such a model, microsystem and apparatus together form a binary scattering system as defined in §4, and are thus to be described in the state space $\underline{H} = H \otimes H_a$. It is convenient here to use the Dirac picture. The independent preparation of microsystem and apparatus means that the incoming state of the composite system is uncorrelated,

$$\underline{W}^{in} = W \otimes W_a .$$

Then the outgoing state is

$$\underline{W}^{out} = \underline{S}(W \otimes W_a)\underline{S}^*$$

with a unitary "scattering" operator $\underline{S}$ on $\underline{H}$ which describes the overall effect of the interaction, and is therefore also fixed by the construction of the apparatus f and the kind of microsystem to which it is applied.

(In actual experiments, an ensemble of N composite systems will not really consist of N copies of the apparatus, each one combined with a single microsystem. Instead one uses a single apparatus f only, which is put again into the state W_a, and combined with another single microsystem in the state W, after the completion of a single measurement. This does not change the formalism, however, since the theory is invariant under time translations, so that single experiments performed at different times can be considered as repetitions of "the same" experiment.)

Separate yes-no measurements on microsystem and apparatus are described, both before and after the interaction, by operators of the form $G \otimes 1_a$ and $1 \otimes G_a$ with $G \in L(H)$ and $G_a \in L(H_a)$, respectively, whereas correlation effects are of the form $G \otimes G_a$. Thus, in particular, the final "reading" of the apparatus f is decribed as the measurement of the particular effect $1 \otimes F_a$ in the state $\underline{W}^{out}$. This measurement gives the result "yes" - i.e., the apparatus f is triggered - with probability

$$f(W) = tr((1 \otimes F_a)\underline{W}^{out}) = tr((1 \otimes F_a)\underline{S}(W \otimes W_a)\underline{S}^*) \ . \qquad (5.1)$$

With W replaced by an arbitrary $T \in B(H)_1$, the last expression defines a linear functional f(T) on $B(H)_1$. This functional is continuous with respect to the trace norm since, along with T and W_a, $T \otimes W_a$ and $\underline{S}(T \otimes W_a)\underline{S}^*$ are also trace class operators, and

$$\|\underline{S}(T \otimes W_a)\underline{S}^*\|_1 = \|T \otimes W_a\|_1 = \|T\|_1 \ ,$$

so that, as $\|1 \otimes F_a\| = \|F_a\| \leq 1$,

$$|f(T)| \leq \|1 \otimes F_a\| \ \|\underline{S}(T \otimes W_a)\underline{S}^*\|_1 \leq \|T\|_1 \ .$$

Therefore f(T) may be rewritten in the form

$$f(T) = tr(FT)$$

with a unique operator $F \in B(H)$. Moreover, since (5.1) implies $0 \leq f(W) = tr(FW) \leq 1$ for all $W \in K(H)$, F satisfies $0 \leq F \leq 1$, and thus belongs to L(H). Since the apparatus f is thus triggered with probability tr(FW) by the microsystems in the (arbitrarily given) state W, it indeed measures a uniquely determined effect F.

The state of the microsystems which leave the apparatus can be determined by measuring on them arbitrary effects $G \in L(H)$ after their interaction with the apparatus. Assume first that no selection is made. Then such measurements are to be described simply as measurements of the subsystem effects $G \otimes 1_a$ in the state $\underline{W}^{out}$ of the composite system. These effects thus occur with the probabilities

$$\tilde{w}(G) = tr((G \otimes 1_a)\underline{W}^{out}) = tr(G\tilde{W}) \ ,$$

where

$$\tilde{W} = Tr_a\underline{W}^{out} = Tr_a(\underline{S}(W \otimes W_a)\underline{S}^*) \qquad (5.2)$$

according to (4.11). Here Tr_a denotes the partial trace with respect to H_a; i.e., $\tilde{W}$ is the reduction of the state $\underline{W}^{out}$ to the microsystem. This shows that the interaction with the apparatus transforms an arbitrary initial state W of the microsystem into a final state $\tilde{W}$ given by (5.2).

The mapping $\tilde{\phi} : W \to \tilde{W}$ defined by (5.2) may be extended immediately, in the form

$$\tilde{\phi}T = Tr_a(\underline{S}(T \otimes W_a)\underline{S}^*) \ , \qquad (5.3)$$

to arbitrary $T \in B(H)_1$. Since one can prove quite easily that (5.3) defines a completely positive linear mapping $\tilde{\phi}$ of $B(H)_1$ into itself, this implies that $\tilde{\phi}$ describes an operation. We postpone this proof until a little later, however. As expected, $\tilde{\phi}$ is a non-selective operation, since (5.2) implies $tr(\tilde{\phi}W) = tr\tilde{W} = tr\underline{W}^{out} = 1$ for all W.

The situation becomes more complicated if one uses the apparatus f to perform a selective operation, and wants to determine the state of the subensemble of those microsystems which have triggered the effect F. Assume one has performed the F measurement $N \gg 1$ times. Then the selected subensemble consists of

$$\hat{N} = N\, tr(FW) = N\, tr((1 \otimes F_a)\underline{W}^{out}) \qquad (5.4)$$

microsystems, according to (5.1). Again the state $\hat{W}$ of this subensemble is to be determined by subsequent measurements of arbitrary effects $G \in L(H)$ on these microsystems. Such measurements give the result "yes" in

$$\hat{N}_+ = N\, tr((G \otimes F_a)\underline{W}^{out}) \qquad (5.5)$$

cases, by defintion of the correlation effects $G \otimes F_a$. Indeed, the occurence of the subsystem effect $G \otimes 1_a$ in the subensemble of composite systems selected according to the subsystem effect $1 \otimes F_a$ means that both $G \otimes 1_a$ and $1 \otimes F_a$ occur together; and both measurements are performed in the state $\underline{W}^{out}$. The probability for the occurrence of the effect G in the selected subensemble of microsystems is thus, by (5.4) and (5.5),

$$\hat{w}(G) = \frac{\hat{N}_+}{\hat{N}} = \frac{tr((G \otimes F_a)\underline{W}^{out})}{tr((1 \otimes F_a)\underline{W}^{out})} . \qquad (5.6)$$

Due to the cyclic interchangeability of operators under the trace, the numerator of the last expression may be rewritten in the form

$$tr((G \otimes 1_a)(1 \otimes F_a^{\frac{1}{2}})\underline{W}^{out}(1 \otimes F_a^{\frac{1}{2}})) = tr(G\cdot\phi W) ,$$

with

$$\phi W = Tr_a((1 \otimes F_a^{\frac{1}{2}})\underline{W}^{out}(1 \otimes F_a^{\frac{1}{2}}))$$

$$= Tr_a((1 \otimes F_a^{\frac{1}{2}})\underline{S}(W \otimes W_a)\underline{S}^*(1 \otimes F_a^{\frac{1}{2}})) , \qquad (5.7)$$

according to (4.12).

Since the mapping Tr_a is trace preserving,

$$tr(\phi W) = tr((1 \otimes F_a^{\frac{1}{2}})\underline{W}^{out}(1 \otimes F_a^{\frac{1}{2}}))$$

$$= tr((1 \otimes F_a)\underline{W}^{out}) \qquad (5.8)$$

coincides with the denominator of the last expression in (5.6), so that we may rewrite this equation in the form

$$\hat{w}(G) = tr(G\hat{W}) , \quad \hat{W} = \phi W/tr(\phi W) ; \qquad (5.9)$$

i.e., the selected subensemble of microsystems is in the state $\hat{W}$. Moreover, by (5.1), (5.8) and the definition of F, $tr(\phi W)$ also coincides with the probability $tr(FW)$ of the measured effect F in the state

W, which is the transition probability $\hat{N}/N$ for the performed selection procedure. (Thus, as discussed in §2, $tr(\phi W) = 0$ again means that the subensemble is empty, and therefore $\hat{W}$ need not - and can not - be defined via (5.9).) We therefore expect that the mapping $\phi : W \to \phi W$ defined by (5.7) describes a selective operation in the sense of §2.

To show this, we first extend this mapping ϕ to $B(H)_1$ in an obvious way, by defining

$$\phi T = Tr_a((1 \otimes F_a^{\frac{1}{2}})\underline{S}(T \otimes W_a)\underline{S}^*(1 \otimes F_a^{\frac{1}{2}})) \qquad (5.10)$$

for arbitrary $T \in B(H)_1$. For such T, the operator

$$(1 \otimes F_a^{\frac{1}{2}})\underline{S}(T \otimes W_a)\underline{S}^*(1 \otimes F_a^{\frac{1}{2}}))$$

belongs to $B(\underline{H})_1$, since $B(\underline{H})_1$ is a two-sided ideal in $B(\underline{H})$ (cf. [3], Ch. 1, or [4]). It depends linearly on T, and is positive if T is. Since Tr_a maps $B(\underline{H})_1$ into $B(H)_1$ and is linear and positive, (5.10) defines a positive linear mapping $\phi : B(H)_1 \to B(H)_1$. Moreover, $tr(\phi W) \leq 1$ for all W, as shown above. Therefore it only remains to prove complete positivity of ϕ.

Take, for this purpose, an arbitrary finite-dimensional or separable Hilbert space $\tilde{H}$, and consider the product space $H \otimes \tilde{H} \otimes H_a$. The tensor product of Hilbert spaces is associative and commutative; i.e., the spaces $(H \otimes H_a) \otimes \tilde{H}$, $(H \otimes \tilde{H}) \otimes H_a$, etc., can all be identified with $H \otimes \tilde{H} \otimes H_a$ in a "natural" and obvious way. Therefore operators like $\underline{S} \otimes \tilde{1}$, or $\mathbf{T} \otimes W_a$ with $\mathbf{T} \in B(H \otimes \tilde{H})_1$ and $W_a \in K(H_a)$, etc., can be considered as operating on $H \otimes \tilde{H} \otimes H_a$. Define now, for arbitrary $\mathbf{T} \in B(H \otimes \tilde{H})_1$,

$$\underline{\phi}\mathbf{T} = \underline{Tr}_a((1 \otimes \tilde{1} \otimes F_a^{\frac{1}{2}})(\underline{S} \otimes \tilde{1})(\mathbf{T} \otimes W_a)(\underline{S} \otimes \tilde{1})^*(1 \otimes \tilde{1} \otimes F_a^{\frac{1}{2}})) , \qquad (5.11)$$

with $\underline{Tr}_a$ denoting the partial trace with respect to H_a which maps

$B(H \otimes \underset{\sim}{H} \otimes H_a)_1$ onto $B(H \otimes \underset{\sim}{H})_1$, and which therefore has to be distinguished from the analogous mapping $Tr_a : B(H \otimes H_a)_1 \to B(H)_1$ considered before. Eq. (5.11) is of the same form as (5.10), with H_a, W_a and F_a unchanged, but with H replaced by $H \otimes \underset{\sim}{H}$ (and thus $H \otimes H_a$ by $H \otimes \underset{\sim}{H} \otimes H_a$, 1 by $1 \otimes \underset{\sim}{1}$, and Tr_a by $\underline{Tr}_a$), T by $\mathbf{T}$, and $\underline{S}$ by $\underline{S} \otimes \underset{\sim}{1}$ (which is also unitary). Therefore (5.11) defines a positive linear mapping $\underline{\phi}$ of $B(H \otimes \underset{\sim}{H})_1$ into itself. With $T \in B(H)_1$ and $\underset{\sim}{T} \in B(\underset{\sim}{H})_1$, we have

$$(\underline{S} \otimes \underset{\sim}{1})(T \otimes \underset{\sim}{T} \otimes W_a)(\underline{S} \otimes \underset{\sim}{1})^* = (\underline{S}(T \otimes W_a)\underline{S}^*) \otimes \underset{\sim}{T} ,$$

so that (5.11) implies

$$\begin{aligned}\underline{\phi}(T \otimes \underset{\sim}{T}) &= \underline{Tr}_a([(1 \otimes F_a^{\frac{1}{2}})\underline{S}(T \otimes W_a)\underline{S}^*(1 \otimes F_a^{\frac{1}{2}})] \otimes \underset{\sim}{T}) \\ &= (Tr_a[(1 \otimes F_a^{\frac{1}{2}})\underline{S}(T \otimes W_a)\underline{S}^*(1 \otimes F_a^{\frac{1}{2}})]) \otimes \underset{\sim}{T} \\ &= \phi T \otimes \underset{\sim}{T} . \end{aligned} \qquad (5.12)$$

Here we have used (5.10) and the fact that, for arbitrary $\underline{R} \in B(H \otimes H_a)_1$ and $\underset{\sim}{T} \in B(\underset{\sim}{H})_1$,

$$\underline{Tr}_a(\underline{R} \otimes \underset{\sim}{T}) = (Tr_a\underline{R}) \otimes \underset{\sim}{T} . \qquad (5.13)$$

(The latter is easily proved, e.g., with the help of suitably generalized versions of Eq. (4.14).)

Taking $\underset{\sim}{H}$ n-dimensional, (5.12) implies that $\underline{\phi}$ coincides with the mapping ϕ_n used in §2 to define n-positivity of ϕ. As $\underline{\phi}$ is positive for all n, this establishes complete positivity of ϕ, whereas for infinite-dimensional $\underset{\sim}{H}$ we obtain the (apparently) stronger positivity property discussed before.

Because Eq. (5.3) results from (5.10) by substituting for F_a the unit operator 1_a, the arguments of the last two paragraphs apply to the mapping $\tilde{\phi}$ as well; i.e., $\tilde{\phi}$ is also linear and completely positive. Thus, indeed, both ϕ and $\tilde{\phi}$ are operations. By rearranging the operators $1 \otimes F_a^{1/2}$ under the partial trace, one obtains two equivalent but simpler versions,

$$\phi T = \mathrm{Tr}_a((1 \otimes F_a)\underline{S}(T \otimes W_a)\underline{S}^*) \ , \tag{5.14}$$

or

$$\phi T = \mathrm{Tr}_a(\underline{S}(T \otimes W_a)\underline{S}^*(1 \otimes F_a)) \ ,$$

of Eq. (5.10). Such rearrangements are possible since, for arbitrary $\underline{T} \in B(\underline{H})_1$, $X_a \in B(H_a)$ and $X \in B(H)$,

$$\mathrm{tr}(X \cdot \mathrm{Tr}_a((1 \otimes X_a)\underline{T})) = \mathrm{tr}((X \otimes 1_a)(1 \otimes X_a)\underline{T})$$

$$= \mathrm{tr}((X \otimes 1_a)\underline{T}(1 \otimes X_a)) = \mathrm{tr}(X \cdot \mathrm{Tr}_a(\underline{T}(1 \otimes X_a))) \ ,$$

and thus

$$\mathrm{Tr}_a((1 \otimes X_a)\underline{T}) = \mathrm{Tr}_a(\underline{T}(1 \otimes X_a)) \ . \tag{5.15}$$

If the apparatus f is used to select the subensemble of microsystems which have not triggered it - i.e., which have not produced the apparatus effect F_a - the corresponding operation ϕ' is obtained simply by replacing F_a by $F'_a = 1_a - F_a$ in (5.14). Then, for arbitrary $T \in B(H)_1$,

$$\phi T + \phi' T = \mathrm{Tr}_a\left[(1 \otimes F_a)\underline{S}(T \otimes W_a)\underline{S}^* + (1 \otimes F'_a)\underline{S}(T \otimes W_a)\underline{S}^*\right]$$

$$= \mathrm{Tr}_a(\underline{S}(T \otimes W_a)\underline{S}^*) = \tilde{\phi} T \ ,$$

and thus, in particular,

$$\mathrm{tr}(\phi W) + \mathrm{tr}(\phi' W) = \mathrm{tr}(\tilde{\phi} W) = 1$$

for all $W \in K(H)$. Therefore ϕ and ϕ' are complementary operations (cf. (2.37)), and $\tilde{\phi}$ is the non-selective operation $\phi + \phi'$ associated with ϕ and ϕ', as expected from the construction of the model.

The adjoint ϕ^* of the mapping ϕ is defined implicitly (cf. (2.25), (5.14) and (4.12)) by

$$\operatorname{tr}(\phi^*X\cdot T) = \operatorname{tr}(X\cdot\phi T) = \operatorname{tr}((X \otimes 1_a)(1 \otimes F_a)\underline{S}(T \otimes W_a)\underline{S}^*)$$

$$= \operatorname{tr}((T \otimes W_a)\underline{S}^*(X \otimes F_a)\underline{S}) \qquad (5.16)$$

for arbitrary $X \in B(H)$ and $T \in B(H)_1$. After inserting $T = |f\rangle\langle f|$ with an arbitrary unit vector $f \in H$, evaluating the first and the last trace in (5.16) with suitable orthogonal bases in H and $H \otimes H_a$, respectively, and exploiting the polarization identity (1.13), we obtain

$$(f,\phi^*Xg) = \sum_k \left((f \otimes g_a^k),(1 \otimes W_a)\underline{S}^*(X \otimes F_a)\underline{S}(g \otimes g_a^k)\right) , \qquad (5.17)$$

valid for all f, $g \in H$ and an arbitrary orthogonal basis $\{g_a^k\}$ in H_a. (Compare the analogous derivation of (4.14) from (4.12).) Comparing this with (4.14), we see that (5.17) can be rewritten formally as

$$\phi^*X = \operatorname{Tr}_a((1 \otimes W_a)\underline{S}^*(X \otimes F_a)\underline{S}) . \qquad (5.18)$$

More precisely, we have proved that the partical trace mapping Tr_a, when defined by (4.14), can be extended to the operators $(1 \otimes W_a)\underline{S}^*(X \otimes F_a)\underline{S}$ (which need not belong to $B(H \otimes H_a)_1$), and that (5.18) holds true with this extension. In particular, the effect $F = \phi^*1$ corresponding to ϕ is given explicitly by

$$F = \operatorname{Tr}_a((1 \otimes W_a)\underline{S}^*(1 \otimes F_a)\underline{S}). \qquad (5.19)$$

Eqs. (5.18) and (5.19) are not very useful in practice, however, and shall thus not be discussed further.

The results obtained so far are summarized and extended by

Theorem 2 (Second Representation Theorem):

Let H_a be a (finite-dimensional, separable or even non-separable) Hilbert space, W_a a statistical operator and F_a an effect operator on H_a, and $\underline{S}$ a unitary operator on $H \otimes H_a$. Then

$$\phi T = \mathrm{Tr}_a((1 \otimes F_a)\underline{S}(T \otimes W_a)\underline{S}^*) \ , \tag{5.20}$$

with $T \in B(H)_1$ arbitrary, defines an operation ϕ. With $F'_a = 1_a - F_a$, the operation ϕ' defined by

$$\phi' T = \mathrm{Tr}_a((1 \otimes F'_a)\underline{S}(T \otimes W_a)\underline{S}^*) \tag{5.21}$$

is complementary to ϕ, and the non-selective operation $\tilde{\phi} = \phi + \phi'$ associated with ϕ and ϕ' is given by

$$\tilde{\phi} T = \mathrm{Tr}_a(\underline{S}(T \otimes W_a)\underline{S}^*) \ . \tag{5.22}$$

Vice versa, given any two complementary operations ϕ and ϕ' on $B(H)_1$, there exist a Hilbert space H_a, operators $W_a \in K(H_a)$, $F_a \in L(H_a)$, and a unitary operator $\underline{S}$ on $H \otimes H_a$, such that ϕ, ϕ' and $\tilde{\phi} = \phi + \phi'$ are represented by Eqs. (5.20), (5.21) and (5.22), respectively. One may also require, in addition, that H_a is separable, W_a a pure state (i.e., a one-dimensional projection operator), and F_a a decision effect (i.e., a projection operator).-

Proof: The first part of the theorem has already been proved. (Note that the dimension of H_a in the above model was arbitrary.) To prove the second - and more interesting - part, we start from the representations (3.19) and (3.20) of ϕ and ϕ',

$$\phi T = \sum_{k \in K} A_k \, T \, A_k^* \ , \qquad \phi' T = \sum_{k \in K'} A_k \, T \, A_k^* \ , \tag{5.23}$$

as provided by Theorem 1', and construct the Hilbert space H_a and the operators W_a, F_a and $\underline{S}$ explicitly. For the sake of definiteness, the index set J, of which the sets K and K' occuring in (5.13) are complementary subsets, is taken to be $J = \{1,2,\ldots\}$. This can be done without loss of generality by setting $A_k = 0$ for $k > N$ if the original index set J was finite, $J = \{1\ldots N\}$.

Take for H_a a separable Hilbert space with orthogonal basis $\{g_i^a \mid i = 0,1,2\ldots\}$. Then $H \otimes H_a = \underline{H}$ may be decomposed, in the form

$$\underline{H} = \bigoplus_{i\geq 0} (H \otimes g_i^a) = \bigoplus_{i\geq 0} H_i \ ,$$

into orthogonal subspaces $H_i = H \otimes g_i^a$ isomorphic to, and therefore identified with, H. Taking the subspace $H_o \equiv H$ apart, we may also write

$$\underline{H} = H \oplus \hat{H} \ , \quad \hat{H} = \bigoplus_{k\geq 1} H_k \ , \quad H_k \equiv H \ .$$

In the following, the indices i and k are always assumed to take the values $i = 0,1,2\ldots$ and $k = 1,2\ldots$, which somewhat simplifies the notation.

Define, in terms of the operators A_k, $k = 1,2\ldots$ entering (5.23), an operator $\underline{A} : H \to \hat{H} = \oplus_k H_k$ by

$$\underline{A}f = \bigoplus_k A_k f \ .$$

(Note that, for arbitrary $f \in H$, $A_k f \in H \equiv H_k$.) Eq. (3.18) of Theorem 1' then implies

$$(\underline{A}f,\underline{A}g) = \sum_k (A_k f, A_k g) = \sum_k (f, A_k^* A_k g) = (f,g) \ ;$$

i.e., $\underline{A}$ is isometric, with operator norm $\|\underline{A}\| = 1$. Its adjoint $\underline{A}^* : \hat{H} \to H$, defined by

$$(f,\underline{A}^*\hat{f}) = (\underline{A}f,\hat{f}) \tag{5.24}$$

for all $f \in H$ and $\hat{f} \in \hat{H}$, is also bounded, with

$$\|\underline{A}^*\| = \operatorname{Sup} \frac{|(f,\underline{A}^*\hat{f})|}{\|f\|\;\|\hat{f}\|} = \operatorname{Sup} \frac{|(\underline{A}f,\hat{f})|}{\|f\|\;\|\hat{f}\|} = \|\underline{A}\| = 1 \ .$$

Isometry of $\underline{A}$ implies

$$\underline{A}^*\underline{A} = 1 \ , \quad \underline{A}\underline{A}^* = \hat{E} \tag{5.25}$$

with some projection operator $\hat{E}$ on $\hat{H}$. Indeed, $\underline{A}^*\underline{A} : H \to H$ and $\underline{A}\underline{A}^* : \hat{H} \to \hat{H}$ are obviously self-adjoint, and $\underline{A}^*\underline{A} = 1$ since

$$(f,\underline{A}^*\underline{A}g) = (\underline{A}f,\underline{A}g) = (f,g) \ ;$$

thus

$$\hat{E}^2 = \underline{A}\underline{A}^*\underline{A}\underline{A}^* = \underline{A}\underline{A}^* = \hat{E} \ .$$

Explicitly, $\underline{A}^*$ is given by

$$\underline{A}^* : \bigoplus_k f_k \to \sum_k A_k^* f_k \tag{5.26}$$

for arbitrary $\oplus_k f_k \in \hat{H}$ (i.e., $f_k \in H$, $\sum_k \|f_k\|^2 < \infty$), since then

$$(\underline{A}^*(\bigoplus_k f_k),f) = (\sum_k A_k^* f_k,f) = \sum_k (f_k,A_k f)$$

$$= (\bigoplus_k f_k, \bigoplus_k A_k f) = (\bigoplus_k f_k, \underline{A}f) \ ;$$

i.e., (5.26) is satisfied.

Vectors $\underline{f} \in \underline{H} = H \oplus \hat{H}$ may be represented in matrix notation as

$$\underline{f} = f \oplus \hat{f} = \begin{pmatrix} f \\ \hat{f} \end{pmatrix}$$

with $f \in H$ and $\hat{f} \in \hat{H}$. In this notation, bounded operators $\underline{X}$ on $\underline{H}$ can be written as operator matrices,

$$\underline{X} = \begin{pmatrix} X_1 & X_2 \\ X_3 & X_4 \end{pmatrix} ,$$

with bounded operators

$$X_1 : H \to H , \quad X_2 : \hat{H} \to H , \quad X_3 : H \to \hat{H} , \quad X_4 : \hat{H} \to \hat{H} ,$$

so that, according to "matrix multiplication",

$$\underline{X}\,\underline{f} = \begin{pmatrix} X_1 f + X_2 \hat{f} \\ X_3 f + X_4 \hat{f} \end{pmatrix} .$$

The adjoint of $\underline{X}$ is easily seen to be

$$\underline{X}^* = \begin{pmatrix} X_1^* & X_3^* \\ X_2^* & X_4^* \end{pmatrix} ,$$

while operator products are to be calculated by "matrix multiplication", preserving the ordering of the (non-commuting) "matrix elements", of the operator matrices.

We now define, in this matrix notation, the operator $\underline{S}$ on $\underline{H} = H \oplus H_a$ by

$$\underline{S} = \begin{pmatrix} 0 & \underline{A}^* \\ \underline{A} & \hat{1} - \underline{\underline{AA^*}} \end{pmatrix} . \tag{5.27}$$

Obviously $\underline{S}^* = \underline{S}$; moreover, $\underline{S}$ is also unitary:

$$\underline{S}^*\underline{S} = \underline{\underline{SS^*}} = \underline{S}^2 = \begin{pmatrix} \underline{A}^*\underline{A} & \underline{A}^*(\hat{1} - \underline{\underline{AA^*}}) \\ (\hat{1} - \underline{\underline{AA^*}})\underline{A} & \underline{\underline{AA^*}} + (\hat{1} - \underline{\underline{AA^*}})^2 \end{pmatrix}$$

$$= \begin{pmatrix} 1 & 0 \\ 0 & \hat{1} \end{pmatrix} = \underline{1} .$$

Here we have used the following facts: By (5.25), $\underline{A}^*\underline{A} = 1$, and $\hat{1} - \underline{\underline{AA^*}} = \hat{1} - \hat{E}$ is a projection operator, so that

$$\underline{A}^*(\hat{1} - \underline{\underline{AA^*}}) = \underline{A}^* - \underline{A}^*\underline{\underline{AA^*}} = \underline{A}^* - \underline{A}^* = 0 ,$$

$$(\hat{1} - \underline{\underline{AA^*}})\underline{A} = \underline{A} - \underline{\underline{AA^*}}\underline{A} = \underline{A} - \underline{A} = 0 ,$$

and

$$\underline{\underline{A}}\underline{\underline{A}}^* + (\hat{1} - \underline{\underline{A}}\underline{\underline{A}}^*)^2 = \underline{\underline{A}}\underline{\underline{A}}^* + \hat{1} - \underline{\underline{A}}\underline{\underline{A}}^* = \hat{1} .$$

For W_a we take the projection operator onto g_o^a,

$$W_a = | g_o^a >< g_o^a | . \tag{5.28}$$

Then we obtain, for arbitrary $T \in B(H)_1$ and $f \in H$,

$$(T \otimes W_a)(f \otimes g_j^a) = \delta_{jo}(Tf \otimes g_o^a) . \tag{5.29}$$

Due to the isomorphism $H \otimes H_a \equiv \oplus_i H_i$, $H_i \equiv H$, the vector $\underline{f}_j = f \otimes g_j^a$ may also be written as $\oplus_i \delta_{ji} f$. In the above matrix notation, this means

$$\underline{f}_o = \begin{pmatrix} f \\ 0 \end{pmatrix} , \quad \underline{f}_j = \begin{pmatrix} 0 \\ \oplus_k \delta_{jk} f \end{pmatrix} \text{ for } j > 0 ,$$

so that (5.29) takes the form

$$(T \otimes W_a)\underline{f}_o = \begin{pmatrix} Tf \\ 0 \end{pmatrix} , \quad (T \otimes W_a)\underline{f}_j = \begin{pmatrix} 0 \\ 0 \end{pmatrix} \text{ for } j > 0 .$$

Since vectors of the form $\underline{f}_j$ span $\underline{H}$, this means that $T \otimes W_a$ has the matrix representation

$$T \otimes W_a = \begin{pmatrix} T & 0 \\ 0 & 0 \end{pmatrix} . \tag{5.30}$$

Define now, for an arbitrary (finite or infinite) subset κ of $J = \{1,2\ldots\}$, a projection operator $\hat{E}_\kappa$ on $\hat{H}$ by

$$\hat{E}_\kappa(\oplus_k f_k) = \oplus_k g_k , \quad g_k = \begin{cases} f_k & \text{if } k \in \kappa \\ 0 & \text{if } k \notin \kappa \end{cases} ,$$

and two projection operators,

$$E_\kappa^a = \sum_{k\in\kappa} | g_k^a >< g_k^a | , \quad E_{o,\kappa}^a = | g_o^a >< g_o^a | + E_\kappa^a ,$$

on H_a. Then, by arguments similar to the ones leading to (5.30), one easily verifies the relations

$$1 \otimes E^a_\kappa = \begin{pmatrix} 0 & 0 \\ 0 & \hat{E}_\kappa \end{pmatrix}, \quad 1 \otimes E^a_{o,\kappa} = \begin{pmatrix} 1 & 0 \\ 0 & \hat{E}_\kappa \end{pmatrix}. \tag{5.31}$$

A straightforward calculation with (5.27), (5.30) and (5.31) then yields

$$\underline{B} \overset{=}{\text{df.}} (1 \otimes E^a_\kappa)\underline{S}(T \otimes W_a)\underline{S}(1 \otimes E^a_\kappa)$$

$$= (1 \otimes E^a_{o,\kappa})\underline{S}(T \otimes W_a)\underline{S}(1 \otimes E^a_{o,\kappa})$$

$$= \begin{pmatrix} 0 & 0 \\ 0 & \hat{E}_\kappa \underline{A} T \underline{A}^* \hat{E}_\kappa \end{pmatrix} \overset{=}{\text{df.}} \begin{pmatrix} 0 & 0 \\ 0 & \hat{B} \end{pmatrix}. \tag{5.32}$$

Since $T \in B(H)_1$, we have $\underline{B} \in B(\underline{H})_1$, and thus $Tr_a \underline{B} \in B(H)_1$. The partial trace can be calculated by (a suitable analogue of) Eq. (4.14) with the particular basis $\{g^a_i\}$ in H_a used before; for arbitrary f, g $\in$ H we then get

$$(f, Tr_a \underline{B}\, g) = \sum_{j>0} ((f \otimes g^a_j), \underline{B}(g \otimes g^a_j)) . \tag{5.33}$$

As above, we have in matrix notation

$$g \otimes g^a_o = \begin{pmatrix} g \\ 0 \end{pmatrix}, \quad g \otimes g^a_j = \begin{pmatrix} 0 \\ \oplus_k \delta_{jk} g \end{pmatrix} \quad \text{for } j > 0 ,$$

and similarly with f instead of g. Thus (5.32) and (5.33) lead to

$$(f, Tr_a \underline{B}\, g) = \sum_{j>0} ((\bigoplus_k \delta_{jk} f), \hat{B}(\bigoplus_k \delta_{jk} g))$$

$$= \sum_{j>0} (\underline{A}^* \hat{E}_\kappa (\bigoplus_k \delta_{jk} f), T\underline{A}^* \hat{E}_\kappa (\bigoplus_k \delta_{jk} g)) .$$

By definition, $\hat{E}_\kappa$ reproduces $\oplus_k \delta_{jk} f$ if $j \in \kappa$, and annihilates it if $j \notin \kappa$. Therefore the operators E_κ may be dropped in the last expression, if the sum over j is restricted to the index set κ. Eq. (5.26) implies

$$\underline{A}^*(\oplus_k \delta_{jk} f) = A_j^* f ,$$

so that, finally,

$$(f, \mathrm{Tr}_a \underline{B}\, g) = \sum_{j\varepsilon\kappa} (A_j^* f, T A_j^* g) = \sum_{j\varepsilon\kappa} (f, A_j T A_j^* g) ;$$

i.e.,

$$\mathrm{Tr}_a \underline{B} = \sum_{j\varepsilon\kappa} A_j\, T\, A_j^* . \qquad (5.34)$$

(Note that, as shown in §3, the sum in (5.34) is convergent also in the case where κ is infinite, since $\sum_k A_k^* A_k = 1$.)

On the other hand, by the definition (5.32) of $\underline{B}$, and since $\underline{S}^* = \underline{S}$, we also have

$$\mathrm{Tr}_a \underline{B} = \mathrm{Tr}_a((1 \otimes E_\kappa^a)\underline{S}(T \otimes W_a)\underline{S}^*)$$

$$= \mathrm{Tr}_a((1 \otimes E_{o,\kappa}^a)\underline{S}(T \otimes W_a)\underline{S}^*) . \qquad (5.35)$$

Here we have also used Eq. (5.15) to bring the last operator $1 \otimes E_\kappa^a$ or $1 \otimes E_{o,\kappa}^a$ under the partial trace to the left, where it can be omitted since E_κ^a and $E_{o,\kappa}^a$ are projection operators. Comparing Eqs. (5.34) and (5.35), we now see that we obtain the required representations (5.20) and (5.21) of ϕ and ϕ' by taking either

$$F_a = E_K^a , \quad F_a' = E_{o,K'}^a ,$$

or

$$F_a = E_{o,K}^a , \quad F_a' = E_{K'}^a .$$

(By (5.35), it does not matter which of these we choose.) Since $K \cup K' = \{1,2,\ldots\}$, we have in both cases

$$F_a + F_a' = \sum_{k\varepsilon K} | g_k^a >< g_k^a | + | g_0^a >< g_0^a | + \sum_{k\varepsilon K'} | g_k^a >< g_k^a |$$
$$= \sum_{i\geq 0} | g_i^a >< g_i^a | = 1_a .$$

As announced in the theorem, the operators F_a constructed here are projection operators, and W_a as defined by (5.28) describes a pure state.

We finally remark that the above contruction, obviously, could also be carried through with a Hilbert space H_a of finite dimension $N + 1$, if the set of operators A_k, $k \varepsilon J$ entering (5.23) is finite, i.e., $J = \{1...N\}$.

The theorem just proved allows some conclusions, which shall be discussed now.

We have shown, first, that our quantum mechanical model of an effect apparatus and its interaction with microsystems indeed describes the measurement of an effect $F \varepsilon L(H)$ and the two complementary operations ϕ and ϕ' associated with this measurement. Conversely, according to the second part of the theorem, any pair of complementary operations ϕ and ϕ' could be "realized", at least "in principle", by a suitable apparatus of the type considered, which interacts with the microsystem in an appropriate way - i.e., formally, by an appropriate choice of the state space H_a, initial state W_a and "pointer" effect F_a of the apparatus, and of the scattering operator $\underline{S}$ on $H \otimes H_a$. This, clearly, does not prove that such an apparatus with the required interaction could be actually constructed in the laboratory. In view of the internal consistency of quantum mechanics it is, nevertheless, quite satisfactory that the simplest possible concrete model of yes-no measurements already reproduces the very general class of operations, as introduced in a much more abstract way in §2.

Besides this, Theorem 2 provides additional support for our assumption that not only projection operators E, but rather all operators $F \varepsilon L(H)$ describe possible yes-no measurements. Indeed, even if one insists in "conventional" quantum mechanics for the model apparatus by postulating that the "pointer" effect F_a is always described by a projection operator, the effect F measured by such an apparatus need not be a projection operator. On the contrary, every effect operator $F \in L(H)$ can also be measured "in principle" by such a "conventional" apparatus, according to the last part of Theorem 2. Moreover, the operations ϕ performable by such model apparatuses are also completely general, so that conceivable additional restrictions on ϕ (as, e.g., the condition $\phi^*F = F$ for measurements of the first kind) can not be justified in this way either. (Likewise, the classes of effects F and operations ϕ described by the model are independent of whether or not only pure states are admitted as initial apparatus states W_a. But a restriction to pure states W_a would not look very natural anyway, in view of the macroscopic nature of the apparatus.)

It is frequently claimed that state changes due to measurements - and, in particular, the "wave packet reduction" formulae (2.8) and (2.9) - can not be explained by quantum mechanics. One argues, for instance, that time evolution in quantum mechanics (as described, e.g., by the Schrödinger equation for the state vectors in the Schrödinger picture) always transforms pure states into pure states, whereas the operation (2.9) transforms most pure states into mixtures. Theorem 2 shows that this reasoning is unfounded. Not only the particular cases (2.8) and (2.9), but in fact every pair of complementary operations ϕ and ϕ' can be reproduced by a suitable quantum mechanical model of the measuring process. In these models, the non-selective operation $\tilde{\phi} = \phi + \phi'$ simply results from the interaction of the microsystem with the apparatus during a certain interval of time. (Note that, by (5.22), $\tilde{\phi}$ is independent of the choice - or even the existence - of the effect F_a describing the "reading" of the apparatus.) During this interaction interval the microsystem is an open system, and thus its time evolution can not be described by the usual formalism, which applies to closed (isolated) systems only. Therefore the argument quoted above is misleading; indeed, as shown here, the time evolution of the composite system apparatus plus microsystem - which is again a closed system, and is described here in the Dirac picture - leads to a state $\underline{W}^{out}$ whose reduction to the microsystem is just the state $\tilde{\phi}W$. We have shown, moreover, that the selective operations ϕ and ϕ' can be also understood in terms of conventional quantum mechanics. Since measurements in, e.g., the state $\hat{W} = \phi W/\mathrm{tr}(\phi W)$ produced by ϕ are always preceded by the reading of the apparatus effect F_a, one is in fact dealing with correlation measurements on a composite system. In our model, the operations ϕ and ϕ' just provide a convenient "shorthand" description of such correlation measurements. This description is obtained, as in the case of $\tilde{\phi}$, when the apparatus variables are eliminated with the help of the partial trace mapping Tr_a. We may thus conclude that, contrary to a wide-spread belief, state changes of the type (2.8) and (2.9) can be perfectly well understood, if quantum mechanics is assumed to be valid also for measuring instruments.

Our model also illustrates in a simple way the physical meaning of complete positivity. An operation ϕ as given by Eq. (5.10) above is completely positive, if the mappings $\underline{\phi}$ of $B(H \otimes \underset{\sim}{H})_1$ into itself described by Eq. (5.11) are positive for arbitrary (and, in particular, for all finite-dimensional) Hilbert spaces $\underset{\sim}{H}$. Comparing now Eqs. (5.10) and (5.11), one immediately realizes that $\underline{\phi}$ may also be interpreted as an operation performed by the model apparatus f, but now acting on a composite microsystem which consists of two noninteracting subsystems I and II with state spaces H and $\underset{\sim}{H}$, respectively, and is thus described in the state space $H \otimes \underset{\sim}{H}$. The model apparatus f is the same in both cases, since the state space H_a, the initial apparatus state W_a and the "pointer" effect F_a are identical in Eqs. (5.10) and (5.11). Moreover, the unitary operator in (5.11) describing the "scattering" between the composite microsystem I + II and the apparatus f is given by $\underline{S} \otimes \underset{\sim}{1}$. Obviously this means that, first, the apparatus interacts only with subsystem I while leaving subsystem II unaffected, and second, that the interaction between subsystem I and the apparatus is independent of whether or not subsystem II is present, since the resulting "scattering" is described in both cases by the same operator $\underline{S}$ on $H \otimes H_a$. This interpretation of Eq. (5.11) in terms of our quantum mechanical model is therefore in full accordance with the physical meaning of the mappings $\underline{\phi}$, as discussed in detail in §2.

In spite of its considerable heuristic value, however, one should not overestimate the physical significance of the model. There are good reasons to doubt that quantum mechanics in its present form is the appropriate theory of macroscopic systems like, e.g., measuring instruments. Quantum mechanics describes a macroscopic body as a composite system consisting of (at least) about 10^{24} atomic subsystems. This is a highly redundant description, since the overwhelming majority of "observables" of such a complex system are neither observed in practice (and perhaps not even observable at all, since their measurement might require "instruments" exceeding the size of the universe), nor really significant for the actual behaviour of the system. The latter should rather be describable in terms of a much smaller set of observables, usually called the "macroscopic" ones, which are relatively easily measurable and have suitable "classical" properties. Although numerous attempts have been made to incorporate such ideas into quantum mechanics, the resulting theories can not yet be considered to yield a fully satisfactory description of macroscopic systems, in spite of some partial successes.

But even if quantum mechanics were literally true for macrosystems, our model would still be oversimplified in many cases of practical interest. For instance, every real measuring instrument contains electrons. If now the microsystem contains electrons, too, or is itself an electron, then the tensor product $H \otimes H_a$ is not the appropriate state space for the combined system, since all state vectors have to be totally antisymmetric with respect to permutations of the electrons. In view of all this, our quantum mechanical model of the measuring process should really be considered as a model only, rather than as a complete and realistic description of actual measurements.

There are also applications of Theorem 2 which do not presuppose the validity of quantum mechanics for macrosystems. In such applications, the system described in the state space H_a is also a microsystem, rather than a macroscopic apparatus f.

Denote by s and a the microsystems with state spaces H and H_a, respectively, and assume that, when put together, they form a binary scattering system s + a. If N pairs s + a are prepared in an uncorrelated incoming state $W \otimes W_a = \underline{W}^{in}$, their state after the scattering is $\underline{W}^{out} = \underline{S}\,\underline{W}^{in}\,\underline{S}^*$ with a unitary scattering operator $\underline{S}$ on $H \otimes H_a$, as above, and thus the N systems s are finally in the state $\tilde{W} = Tr_a \underline{W}^{out}$. Keeping the initial state W_a of system a fixed while varying W, this procedure yields again a nonselective operation $\tilde{\phi} : W \to \tilde{W}$ for the system s, as described by Eq. (5.22) of Theorem 2. Moreover, the N systems s can also be separated into two complementary ensembles, if they are selected with respect to the outcomes "yes" or "no", respectively, of a yes-no measurement F_a performed at subsystem a of each pair s + a after the scattering. One then concludes as above that these two subensembles are in the states $\hat{W} = \phi W/tr(\phi W)$ and $\hat{W}' = \phi' W/tr(\phi' W)$, with ϕ and ϕ' given by Eqs. (5.20) and (5.21) of Theorem 2, and consist of $N_+ = tr(\phi W) \cdot N$ and $N_- = tr(\phi' W) \cdot N$ systems, respectively.

An "indirect" measurement of this type, therefore, also yields two complementary operations ϕ and ϕ' and a measurement of the corresponding effect $F = \phi^* 1$ on system s. The combination of a preparing instrument w_a for systems a in the state W_a, a single system a prepared by w_a, and an effect apparatus f_a measuring the effect F_a on system a but insensitive to system s, may be considered together as a single effect apparatus f, to be applied to a single system s in the following way: release the system a from w_a, let it be scattered at the system s, then apply the apparatus f_a to a, and read on f_a the outcome "yes" or "no" of the experiment. Then, as shown above, this composite apparatus f measures the effect F, and may be used to perform the operations ϕ, ϕ' and $\tilde{\phi} = \phi + \phi'$ at system s. Again the effect F will in general not be described by a projection operator, regardless of whether or not this is true for F_a. Since such indirect measurements are not uncommon in practice, this shows once more that a restriction to projection operators as describing yes-no measurements is unnatural, and may in fact lead to internal inconsistencies of the theory.

It has even been argued ([13], Ch. 11) that typical measuring instruments always contain a suitably prepared microsystem a (a "trigger"), which first interacts with the observed system s, and afterwards eventually triggers some observable change on the remaining macroscopic part of the apparatus (the "amplifier"). Taking this for granted, we could describe the interaction between a and s as a scattering process, and consider the "amplifier" as an effect apparatus f_a measuring a certain effect F_a on the "trigger" a, thereby arriving at exactly the situation considered above. Whether or not such a description of the quantum mechanical measuring process is sufficiently realistic and really helpful for a deeper understanding, shall not be discussed here, however.

§6 Coexistent Effects and Observables

A set of effects, $C \subset L(H)$, is called coexistent - or: a set of coexistent effects - if all effects $F \in C$ can be measured together by applying a suitable apparatus to single microsystems. Such an apparatus - abbreviated here by the label c - may be visualized as having several "output channels", one for each effect $F \in C$, whose outputs are either "yes" or "no", and which respond with the appropriate relative frequencies to ensembles of microsystems; i.e., if the apparatus c is applied successively to $N \gg 1$ systems in a state W, then the output channel corresponding to the effect $F \in C$ gives the output "yes" in tr(FW)N cases. The apparatus c may thus be considered as a combination of effect apparatuses f, which perform the joint measurement of all effects $F \in C$ when the combined apparatus c is applied to a single microsystem.

In conventional quantum mechanics only decision effects (projection operators) are considered, and coexistent sets of decision effects are usually called commensurable. (Ludwig [2] defines "commensurability" with a slightly narrower meaning; finally, however, this turns out to be equivalent to "coexistence".) It is one of the most characteristic features of quantum theory, as compared to classical theories, that there are sets of decision effects which are not commensurable. We want to investigate now the corresponding problem for arbitrary effects. The notion of "coexistence" as defined above is due to Ludwig [2].

Commensurability in quantum mechanics is often considered - expressis verbis, or at least tentatively - as synonymous with "simultaneous" measurability. Nothing of this sort is implied, however, by the above definition of coexistence; neither the interactions of the microsystem with all parts of the "composite" apparatus c, nor the responses of the different output channels need be "simultaneous" in any sense.

This may be illustrated by a simple example. Consider two effect apparatuses f and g which, when applied separately, would measure the effects F and G, respectively. Assume that the measurement with the apparatus f is performed earlier than the measurement with the apparatus g. (Note that the times of application are included in the specification of the effect apparatuses f and g. What actually is relevant here are the time intervals I_f and I_g during which the microsystem interacts with the apparatuses f and g, respectively; we assume that the whole interval I_f is earlier than I_g.) Assume also that the apparatus f acts non-destructively, and that the apparatuses f and g do not occupy the same region in space. Then both apparatuses f and g can be put together, and applied successively - first f, than g - to single microsystems. The combination f + g may be considered as a single apparatus c. Having, by construction, two different "output channels", this apparatus c is expected to measure two coexistent effects F_1 and F_2.

These effects F_1 and F_2 are easily determined. To do this, we calculate the probabilities for the triggering of the two "output channels" - i.e., of the parts f and g of the apparatus c - by microsystems in an arbitrary state W. When c is applied successively to N >> 1 systems in this state, these systems first interact with the apparatus f, triggering it in tr(FW)N cases. After this, with $\tilde{\phi}$ denoting the nonselective operation performed by f, the N systems are in the new state $\tilde{W} = \tilde{\phi}W$, therefore triggering the apparatus g in $tr(G\tilde{W})N = tr(G \cdot \tilde{\phi}W)N = tr(\phi^{*}G \cdot W)N$ cases. Since, therefore, the probabilities in question are tr(FW) and $tr(\phi^{*}G \cdot W)$, the effects measured together by c are

$$F_1 = F \ , \quad F_2 = \tilde{\phi}^{*}G \ . \tag{6.1}$$

According to our definition, these two effects are coexistent - but, obviously, they are not measured "simultaneously".

If, in particular, both F and G are decision effects,

$$F = E \ , \quad G = \tilde{E} \ ,$$

and the apparatus f performs an "ideal" measurement (cf. (2.9)),

$$\tilde{\phi}W = EWE + E'WE'$$

with $E' = 1 - E$, then (6.1) reads

$$F_1 = E \ , \quad F_2 = E\tilde{E}E + E'\tilde{E}E' \ . \tag{6.2}$$

In this case F_2 is also a decision effect (and equal to $\tilde{E}$) if and only if E and $\tilde{E}$ commute. Namely, if $[E,\tilde{E}] = 0$, we also have $[E',\tilde{E}] = 0$, and thus

$$F_2 = E^2\tilde{E} + E'^2\tilde{E} = (E + E')\tilde{E} = \tilde{E} \ .$$

Vice versa, if F_2 is a projection operator, the same is true for $H = F_2E = E\tilde{E}E$ since, by (6.2), F_2 commutes with E, and the product of commuting projection operators is again a projection operator. The operator

$$A = \tilde{E}E - E\tilde{E}E$$

then satisfies

$$A^*A = E\tilde{E}E - E\tilde{E}E\tilde{E}E = H - H^2 = 0 \ ,$$

which finally implies

$$0 = A = A^* = A^* - A = [E,\tilde{E}] \ .$$

As a "physical" example, consider two counters, occupying the spatial volumes V_1 and V_2 and operating at times t_1 and $t_2 > t_1$, respectively, therefore measuring - when applied separately - the characteristic functions

$$E = \chi_{V_1}(\mathbf{X}(t_1)) \ , \quad \tilde{E} = \chi_{V_2}(\mathbf{X}(t_2))$$

of the position operators $\mathbf{X}(t)$ at $t = t_1$ and $t = t_2$. But position operators at different times, and thus also E and $\tilde{E}$, do not commute. When applied after the first one, therefore, the second counter measures an effect F_2 which is not a decision effect. (Our assumption that the counters perform instantaneous and "ideal" measurements will not be satisfied by actual counters, however. A more realistic description of counters - if possible at all - would be much more involved, and most likely would show that even a single counter does not measure a decision effect.)

On the other hand, if two projection operators E and $\tilde{E}$ commute, they can be measured together - at least "in principle" - by an apparatus of the type considered, and therefore should describe commensurable decision effects. Indeed, commutativity is well known to be neccessary and sufficient for the commensurability of decision effects in quantum mechanics. A complete and rigorous proof of this criterion will be presented below.

In order to obtain a general coexistence criterion, we first consider the case of two effects, F_1 and F_2. They are coexistent if and only if there exists an apparatus c yielding, when applied to a single microsystem, two outputs - both either "yes" or "no" - which can be considered as results of measurements of F_1 and F_2. For the sake of brevity, these two outputs are simply called 1 and 2 here.

By adding some wiring and electronics, the apparatus c can be modified to yield additional yes-no outputs besides 1 and 2; e.g., an output 1' = "not 1" which is "yes" if output 1 is "no", and vice versa; an output $1 \wedge 2$ = "1 and 2" which is "yes" if and only if both outputs 1 and 2 are "yes"; and an output $1 \vee 2$ = "1 or 2" which is "yes" if and only if at least one of the outputs 1 and 2 is "yes". The technical realization of the corresponding new output channels in terms of the already existing channels for the outputs 1 and 2 is well known to every experimentalist. This procedure can be further continued, thereby leading also to new outputs which look somewhat more complicated when expressed in terms of the original outputs 1 and 2.

As is also well known (and obvious from the given examples), the operations with outputs characterized by the words "not", "and" and "or" satisfy the rules of ordinary logic, i.e., the calculational rules of a Boolean algebra. (Therefore we have already introduced the usual symbols ' for "not", $\wedge$ for "and" and $\vee$ for "or".) With α, β, γ, etc. denoting arbitrary outputs, the most important calculational rules are:

$$\left.\begin{array}{l}\alpha \wedge \alpha = \alpha \ , \quad \alpha \wedge \beta = \beta \wedge \alpha \ , \quad (\alpha \wedge \beta) \wedge \gamma = \alpha \wedge (\beta \wedge \gamma) \\ \text{and analogous rules with } \vee \text{ for } \wedge \ ; \\ (\alpha')' = \alpha \ , \quad (\alpha \vee \beta)' = \alpha' \wedge \beta' \ , \quad (\alpha \wedge \beta)' = \alpha' \vee \beta' \ , \\ \alpha \wedge \alpha' = \emptyset \ , \quad \alpha \vee \alpha' = I \ , \quad \alpha \wedge I = \alpha \vee \emptyset = \alpha \ ; \end{array}\right\} \quad (6.3)$$

and

$$\left.\begin{array}{l}\alpha \wedge (\beta \vee \gamma) = (\alpha \wedge \beta) \vee (\alpha \wedge \gamma) \ , \\ \alpha \vee (\beta \wedge \gamma) = (\alpha \vee \beta) \wedge (\alpha \vee \gamma) \end{array}\right\} \quad (6.4)$$

(the distributive laws). In (6.3) there occur the two "trivial" outputs: I, which is always "yes", and $\emptyset = I'$, which is always "no". Corresponding output channels are also easily incorporated in the given apparatus.

Starting from the original apparatus c with only two output channels, well-known technical manipulations as symbolized by ' , $\wedge$ and $\vee$ thus lead to modified apparatuses with additional output channels. This "enlargement" of the apparatus, when pursued far enough, comes to a "natural" end, however, leading finally to an apparatus b which, in each single experiment, yields the 16 outputs

$$1, \quad 2, \quad 1', \quad 2', \tag{6.5}$$

$$1 \wedge 2, \quad 1 \wedge 2', \quad 1' \wedge 2, \quad 1' \wedge 2', \tag{6.6}$$

$$(1 \wedge 2) \vee (1' \wedge 2'), \quad (1 \wedge 2') \vee (1' \wedge 2), \tag{6.7}$$

$$1 \vee 2, \quad 1 \vee 2', \quad 1' \vee 2, \quad 1' \vee 2', \tag{6.8}$$

$$I, \quad \emptyset. \tag{6.9}$$

By using the calculational rules (6.3) and (6.4) one easily checks that any further application of the operations $'$, $\wedge$ and $\vee$ to these 16 outputs does not enlarge the given list, so that a further enlargement of the apparatus would not really yield new outputs, but merely duplicate already realized ones. (For instance,

$$(1 \wedge 2)' = 1' \vee 2',$$

$$1 \wedge ((1 \wedge 2') \vee (1' \wedge 2)) = (1 \wedge (1 \wedge 2')) \vee (1 \wedge (1' \wedge 2))$$

$$= (1 \wedge 2') \vee \emptyset = 1 \wedge 2',$$

$$(1' \wedge 2) \vee (1' \wedge 2') = 1' \wedge (2 \vee 2') = 1' \wedge I = 1',$$

etc.) Being thus closed with respect to the operations $'$, $\wedge$ and $\vee$, the set $\underline{B}$ of the 16 outputs (6.5) to (6.9) is a Boolean algebra. As all outputs $\alpha \in \underline{B}$ are obtained by applying these operations to the outputs 1 and 2, $\underline{B}$ is in fact the smallest Boolean algebra containing the outputs 1 and 2 - i.e., the Boolean algebra <u>generated</u> by them.

The 16 output channels corresponding to the outputs (6.5) to (6.9) look different from each other when realized technically by connecting the original outputs 1 and 2 with suitable (e.g., electronic) devices. Therefore we treat here the outputs themselves also as different. In particular cases it may happen, nevertheless, that some of these outputs coincide (i.e., are either all "yes" or all "no") in every single experiment. For instance, the effects $F_1 = F$ and $F_2 = 1 - F = F'$ are coexistent for every $F \in L(H)$: If an effect apparatus f with yes-no output α measures F, then the output α' = "not α" of the same apparatus may be used to measure F'. A corresponding second output channel may thus be added to the apparatus f, so that f becomes an apparatus c yielding two outputs, $1 = \alpha$ for $F_1 = F$, and $2 = \alpha'$ for $F_2 = F'$. This apparatus c may then be further enlarged, as described above, to an apparatus b with 16 different output channels. It is obvious, however, that the outputs 1' and 2 of this apparatus b always coincide. In such cases, one might be tempted to identify coinciding outputs, thereby arriving at smaller Boolean "output" algebra, which may also be realized technically by an apparatus b with less than 16 different output channels. (In the example just considered, the identification 1' = 2 reduces the list (6.5) to (6.9) to four different outputs, 1, 1', I and $\emptyset$, so that only the two "trivial" output channels have to be added to the apparatus c.) Our decision to treat all outputs (6.5) to (6.9) as formally different has, however, the advantage of yielding the same "universal" Boolean algebra $\underline{B}$ also in such exceptional cases.

Two outputs α and β satisfying $\alpha \wedge \beta = \emptyset$ are never both "yes" in a single experiment, and therefore are said to exclude each other. (E.g., the four outputs (6.6) mutually exclude each other.) In this case, $\alpha \vee \beta$ actually means "either α or β", and shall be written $\alpha \dot{\vee} \beta$ when we want to stress this narrower meaning of "or". (We shall, however, not use this notation $\alpha \dot{\vee} \beta$ for "either α or β" $= (\alpha \vee \beta) \wedge (\alpha' \vee \beta') = (\alpha \wedge \beta') \vee (\alpha' \wedge \beta)$ in cases with $\alpha \wedge \beta \neq \emptyset$. Such an output, with $\alpha = 1$ and $\beta = 2$, is listed above under (6.7).) More generally, an output of the form $\alpha_1 \vee \alpha_2 \vee \ldots \vee \alpha_n$, with $\alpha_1 \ldots \alpha_n$ pairwise excluding each other, is also written $\alpha_1 \dot{\vee} \alpha_2 \dot{\vee} \ldots \dot{\vee} \alpha_n$, and is called the disjoint union of the effects $\alpha_1 \ldots \alpha_n$. This generalization from $n = 2$ to $n > 2$ is natural since, by (6.4),

$$\alpha_1 \wedge (\alpha_2 \vee \ldots \vee \alpha_n) = (\alpha_1 \wedge \alpha_2) \vee \ldots \vee (\alpha_1 \wedge \alpha_n) = \emptyset ,$$

and thus

$$\alpha_1 \vee (\alpha_2 \vee \ldots \vee \alpha_n) = \alpha_1 \dot{\vee} (\alpha_2 \vee \ldots \vee \alpha_n) , \text{ etc.}$$

When an apparatus b of the type considered has been applied to a single microsystem, each one of its 16 outputs $\alpha \in \underline{B}$ can be considered as the result of a yes-no measurement - i.e., the result of measuring a certain effect $F_\alpha \in L(H)$ - on the given microsystem. The apparatus b thus measures together all these effects F_α, $\alpha \in \underline{B}$, which form a certain subset B of L(H). As $\alpha \neq \beta$ does not necessarily imply $F_\alpha \neq F_\beta$, B contains at most 16 different effects, but possibly fewer. (In particular, two coinciding outputs - as in the example discussed above - clearly correspond to the same effect. But even if two outputs do not coincide in every single experiment, the corresponding effects may nevertheless be equal, as we shall see later on.) The set B of effects is coexistent, by definition, and it contains the original coexistent effects F_1 and F_2 corresponding to the particular outputs 1 and 2, respectively. Therefore we call B a coexistent completion of the original coexistent set of effects $\{F_1, F_2\}$; the apparatus b is said to realize this coexistent completion B of $\{F_1, F_2\}$.

Although the joint measurement of all effects $F_\alpha \in B$ is most easily visualized - and therefore has been discussed here - in terms of such an "enlarged" apparatus b, the actual construction of this apparatus b is not really necessary for this purpose. The "trivial" outputs I and $\emptyset$ are already fixed prior to any measurement, and all "nontrivial" outputs $\alpha \in \underline{B}$, as listed above under (6.5) to (6.8), can be calculated from the two outputs 1 and 2 of the original apparatus c. (For instance, if the application of this apparatus c to a single microsystem yields "yes" for output 1 and "no" for output 2, the outputs $1 \wedge 2$ and $1' \wedge 2'$ in (6.6) are "no", therefore the output $(1 \wedge 2) \vee (1' \wedge 2')$ in (6.7) is also "no", etc.) Therefore the original apparatus c already measures, at least implicitly, all effects $F_\alpha \in B$.

The correspondence between outputs $\alpha \in \underline{B}$ and effects $F_\alpha \in B$ defines a mapping $\mathbf{F} : \alpha \rightarrow F_\alpha$ of $\underline{B}$ onto $B \subset L(H)$. The properties of this mapping $\mathbf{F}$ are immediately obvious from its physical meaning. First, obviously, the trivial outputs I and $\emptyset$ correspond to the trivial effects represented by the operators 1 and 0, respectively:

$$F_I = 1 , \quad F_\emptyset = 0 . \tag{6.10}$$

Second, for mutually exclusive outputs α and β the probabilities for $\alpha \dot{\vee} \beta$ = "either α or β" to be "yes" must behave additively; i.e.,

$$\mathrm{tr}(F_{\alpha \dot{\vee} \beta} W) = \mathrm{tr}(F_\alpha W) + \mathrm{tr}(F_\beta W)$$

for arbitrary states W, which implies

$$F_{\alpha \dot{\vee} \beta} = F_\alpha + F_\beta . \tag{6.11}$$

This can be easily generalized to disjoint unions of $n > 2$ outputs in the form

$$F_{\alpha_1 \dot{\vee} \ldots \dot{\vee} \alpha_n} = F_{\alpha_1} + \ldots + F_{\alpha_n} . \tag{6.12}$$

Finally, since $\alpha \wedge \alpha' = \emptyset$ and $\alpha \dot{\vee} \alpha' = I$ for all α, (6.10) and (6.11) imply

$$F_{\alpha'} = 1 - F_\alpha = F'_\alpha ; \tag{6.13}$$

i.e., $F_{\alpha'}$ coincides with the effect F'_α = "not F_α" introduced in §2, as expected.

For arbitrary outputs α and β, (6.12) and the relations

$$\alpha = (\alpha \wedge \beta) \dot{\vee} (\alpha \wedge \beta') , \quad \beta = (\alpha \wedge \beta) \dot{\vee} (\alpha' \wedge \beta)$$

and

$$\alpha \vee \beta = (\alpha \wedge \beta) \dot{\vee} (\alpha \wedge \beta') \dot{\vee} (\alpha' \wedge \beta)$$

imply a generalization of (6.11),

$$F_{\alpha \vee \beta} = F_\alpha + F_\beta - F_{\alpha \wedge \beta} . \tag{6.14}$$

In particular, we get from this

$$F_{\alpha \vee \beta} = F_\alpha + F_\beta \quad \text{if} \quad F_{\alpha \wedge \beta} = 0 \; . \tag{6.15}$$

Since $F_{\alpha \wedge \beta} = 0$ means that the outputs α and β are never found to be both "yes", (6.15) has the same physical background as (6.11), but (6.15) is somewhat more general, because $\alpha \wedge \beta = \emptyset$ implies $F_{\alpha \wedge \beta} = 0$ but not vice versa. We leave it to the reader to derive a generalized version of (6.15),

$$F_{\alpha_1 \vee \ldots \vee \alpha_n} = F_{\alpha_1} + \ldots + F_{\alpha_n} \quad \text{if} \quad F_{\alpha_i \wedge \alpha_k} = 0 \quad \text{for } i \neq k \; .$$

Since all properties of the mapping $\mathbf{F} : \alpha \to F_\alpha$ listed so far follow from (6.10) and (6.12), these two requirements are sufficient to characterize $\mathbf{F}$. (Actually, (6.11) and one of the two equations (6.10) would also be sufficient.)

The physical meaning of the effects $F_\alpha \in B$ follows immediately from the interpretation of the corresponding apparatus outputs $\alpha \in \underline{B}$. For instance, the effect $F_{1 \wedge 2}$ is triggered on the apparatus b if and only if both F_1 and F_2 are triggered, and may therefore be called "F_1 and F_2". (We have already used this terminology, for the particular case of subsystem effects, in §4.) Similarly, e.g., $F_{1'}$, $F_{1 \vee 2}$, and $F_{(1 \wedge 2') \vee (1' \wedge 2)}$ may be called "not F_1", "F_1 or F_2", and "either F_1 or F_2", respectively. This notation should not be misunderstood, however, to mean that "not", "and" and "or" as used here represent well-defined calculational rules for effect operators, which would permit to calculate the effects $F_{\alpha'}$ = "not F_α", $F_{\alpha \wedge \beta}$ = "F_α and F_β" and $F_{\alpha \vee \beta}$ = "F_α or F_β" directly and uniquely from the operators F_α and F_β. This is only true for "not", according to (6.13). If there were corresponding rules also for "and" and "or", their successive application - together with (6.13) - would allow one to calculate uniquely all operators $F_\alpha \in B$ from F_1 and F_2. (In fact the existence of one additional rule would suffice, since (6.14) already allows one to calculate $F_{\alpha \vee \beta}$ from $F_{\alpha \wedge \beta}$ and vice versa.) However, as we shall prove later on, the effects F_1 and F_2 do not in general uniquely determine the remaining effects $F_\alpha \in B$. Thus neither "and" nor "or", when used as above, represent unique calculational rules for (coexistent) effect operators.

The preceding discussion yields the following necessary condition for the coexistence of two effects F_1 and F_2:

There exists a mapping $\mathbf{F} : \alpha \in \underline{B} \rightarrow F_\alpha \in B$, satisfying (6.10) and (6.12), of the Boolean algebra $\underline{B}$ generated by two elements 1 and 2 (cf. (6.5) to (6.9)) onto a set B of effects, such that B contains the effects F_1 and F_2 as images under $\mathbf{F}$ of the particular elements 1 and 2 of $\underline{B}$, respectively. (6.16)

The Boolean algebra $\underline{B}$ is considered here as the abstract algebra generated by two elements 1 and 2, its internal structure being indeed independent of any more specific interpretation of its elements. Condition (6.16) is thus a purely mathematical one, not referring any more - as the original definition of coexistence did - to the existence of a measuring apparatus for the joint measurement of F_1 and F_2. Nevertheless, condition (6.16) can be interpreted to mean that one can at least imagine the existence of such an apparatus, which - when suitably extended - has the elements $\alpha \in \underline{B}$ as outputs, and measures together all effects $F_\alpha \in B$. We go beyond this interpretation by assuming condition (6.16) to imply that such an apparatus b can not only be imagined but really constructed (at least "in principle", as theorists usually add). We thus consider condition (6.16) also as sufficient for the coexistence of F_1 and F_2. The set of effects B occuring in (6.16) may then be interpreted as a coexistent completion of the set $\{F_1, F_2\}$, as realized by the apparatus b.

The choice of (6.16) as a necessary and sufficient coexistence condition is motivated by the following arguments. First, no other condition has ever been proposed. Second, (6.16) has a simple physical background, and its mathematical form can also be simplified considerably (see below), so that it is simply applicable as well. Third, if (6.16) is satisfied, one may explicitly construct a quantum mechanical model of an apparatus b for the joint measurement of F_1, F_2 and all other effects $F_\alpha \in B$. Last but not least, when applied to decision effects, (6.16) reproduces the well-established commutativity criterion of conventional quantum mechanics. Before discussing such specific applications of (6.16), however, we shall first transform it into an equivalent but much simpler conditon.

In this connection the four particular elements

$$1 \wedge 2, \quad 1 \wedge 2', \quad 1' \wedge 2, \quad 1' \wedge 2'$$

of $\underline{B}$, listed above under (6.6) and excluding each other pairwise, play a decisive rôle. Every element $\alpha \in \underline{B}$ can be represented in a unique way as a disjoint union of $n \leq 4$ elements of this particular kind: $n = 0$ yields $\alpha = \emptyset$; $n = 1$ yields the four elements (6.6) themselves; for $n = 2$ we get the two elements (6.7) and the four elements (6.5) - the latter, because

$$\begin{aligned} (1 \wedge 2) \dot{\vee} (1 \wedge 2') &= 1 \wedge (2 \vee 2') = 1 \wedge I = 1 \,, \\ (1 \wedge 2) \dot{\vee} (1' \wedge 2) &= (1 \vee 1') \wedge 2 = I \wedge 2 = 2 \,, \end{aligned} \tag{6.17}$$

etc.; for $n = 4$ we get

$$\begin{aligned} &(1 \wedge 2) \dot{\vee} (1 \wedge 2') \dot{\vee} (1' \wedge 2) \dot{\vee} (1' \wedge 2') \\ &= (1 \wedge (2 \vee 2')) \vee (1' \wedge (2 \vee 2')) = 1 \vee 1' = I \,, \end{aligned} \tag{6.18}$$

so that, finally,

$$(1 \wedge 2) \dot{\vee} (1 \wedge 2') \dot{\vee} (1' \wedge 2) = (1' \wedge 2')' = 1 \vee 2 \,,$$

etc.; i.e., $n = 3$ yields the remaining four elements (6.8) of $\underline{B}$. (In the last equation, we have used (6.18) and the fact that $\alpha \dot{\vee} \beta = I$ implies $\beta = \alpha'$.)

Consider two effects F_1 and F_2 satisfying the coexistence condition (6.16). Then there exist four effects,

$$\left.\begin{array}{ll} F_{12} = F_{1\wedge 2} , & F_{12'} = F_{1\wedge 2'} , \\ F_{1'2} = F_{1'\wedge 2} , & F_{1'2'} = F_{1'\wedge 2'} , \end{array}\right\} \tag{6.19}$$

the images under $\mathbf{F}$ of the four elements (6.6) of $\underline{B}$. Eqs. (6.17) and (6.18), together with (6.10) and the additivity property (6.12) of the mapping $\mathbf{F}$, imply

$$F_{12} + F_{12'} = F_1 , \quad F_{12} + F_{1'2} = F_2 , \tag{6.20}$$

and

$$F_{12} + F_{12'} + F_{1'2} + F_{1'2'} = 1 . \tag{6.21}$$

More generally, since an arbitrary element $\alpha \in \underline{B}$ is a disjoint union of $n \leq 4$ elements β_i from (6.6), (6.12) implies that the corresponding effect F_α is the sum of the n effects F_{β_i} from (6.19). The mapping $\mathbf{F} : \underline{B} \to B$ is thus completely specified if the four effects (6.19) are known. Actually it suffices to know three of them, e.g., F_{12}, $F_{12'}$ and $F_{1'2}$. They satisfy, because of (6.21) and $F_{1'2'} \geq 0$,

$$F_{12} + F_{12'} + F_{1'2} \leq 1 , \tag{6.22}$$

and the missing fourth effect $F_{1'2'}$ is obtained from (6.21):

$$F_{1'2'} = 1 - (F_{12} + F_{12'} + F_{1'2}) . \tag{6.23}$$

(Eq. (6.23) immediately implies $F_{1'2'} \leq 1$ and - with (6.22) - also $F_{1'2'} \geq 0$; thus it really defines an effect.)

We have thus shown that, if two effects F_1 and F_2 are coexistent according to condition (6.16), there exist three effects F_{12}, $F_{12'}$ and $F_{1'2}$ satisfying (6.22), such that F_1 and F_2 may be written in terms of these three effects in the form (6.20). Assuming that, conversely, two effects F_1 and F_2 can be represented in the form (6.20) with three effects F_{12}, $F_{12'}$ and $F_{1'2}$ satisfying (6.22), we shall now prove that F_1 and F_2 satisfy the coexistence condition (6.16).

For this purpose, we first define a fourth effect $F_{1'2'}$ by (6.23), so that (6.21) is satisfied. Then we interpret the four effects F_{12} to $F_{1'2'}$, in accordance with (6.19), as images of the four particular elements (6.6) of $\underline{B}$ under a mapping $\mathbf{F} : \alpha \in \underline{B} \to F_\alpha \in L(H)$ of the type considered in condition (6.16). If such a mapping $\mathbf{F}$ exists, then - as shown above - the images F_α under $\mathbf{F}$ of the remaining elements $\alpha \in \underline{B}$ may be represented as sums of the four particular effects (6.19). These representations can now be taken to define F_α for arbitrary $\alpha \in \underline{B}$, and it only remains to show that the mapping $\mathbf{F} : \alpha \to F_\alpha$ obtained in this way has the required properties. As sums of $n \geq 4$ different operators from (6.19), all operators F_α are ≥ 0, and also ≤ 1 by (6.21); therefore they belong to L(H). The validity of the first of Eqs. (6.10) follows from (6.18) and (6.21), whereas the second one is trivially satisfied. The additivity condition (6.12) is an immediate consequence of the explicit construction of the effects F_α. Finally, by (6.17) and (6.20), $\mathbf{F}$ also maps the elements 1 and 2 of $\underline{B}$ into the effects F_1 and F_2, respectively. Thus, indeed, condition (6.16) is satisfied for F_1 and F_2.

We have thus proved that condition (6.16) is equivalent to the following simpler coexistence criterion:

> Two effects F_1 and F_2 are coexistent if and only if they can be represented in the form (cf. (6.20))
>
> $$F_1 = F_{12} + F_{12'} \,, \quad F_2 = F_{12} + F_{1'2}$$
>
> in terms of three effects F_{12}, $F_{12'}$ and $F_{1'2}$ satisfying (cf. (6.22))
>
> $$F_{12} + F_{12'} + F_{1'2} \leq 1 \,. \qquad (6.24)$$

The physical interpretation of the four effects F_{12}, $F_{12'}$, $F_{1'2}$ and $F_{1'2'}$ (the latter being defined by (6.23)) follows immediately from (6.19), and is most easily expressed in the terminology introduced above, according to which they may be called "F_1 and F_2", "F_1 and not F_2", "F_2 and not F_1", and "not F_1 and not F_2" = "neither F_1 nor F_2", respectively. If, therefore, F_1 and F_2 are measured together on single microsystems by a suitable apparatus, then the probability that single microsystems in state W trigger both F_1 and F_2 is $tr(F_{12}W)$, wheras the probability for triggering F_1 but not F_2 is $tr(F_{12'}W)$, etc. In this sense, the operators F_{12} to $F_{1'2'}$ describe the <u>correlations</u> between the results of joint measurements of the two coexistent effects F_1 and F_2. Generalizing a terminology already used in §4, we therefore call them correlation effects.

The three correlation effects F_{12}, $F_{12'}$ and $F_{1'2}$, as shown above, uniquely determine all effects F_α, $\alpha \in \underline{B}$, and thus, among them, also the two effects F_1 and F_2 (cf. (6.20)). If, as usual, the effects F_1 and F_2 are given in advance, it therefore also suffices to know besides them only a single one of these three correlation effects, as the two others may then be calculated from (6.20). Since Eqs. (6.20) and (6.21) also imply

$$F_1' = F_{1'2} + F_{1'2'} \,, \qquad F_2' = F_{12'} + F_{1'2'} \,, \tag{6.25}$$

the knowledge of F_1, F_2 and $F_{1'2'}$ is also sufficient to calculate all effects F_α.

Before drawing general conclusions, we shall first illustrate the coexistence criterion (6.24) by the example (6.1), the successive application of two effect aparatuses f and g. With ϕ and $\tilde{\phi} = \phi + \phi'$ denoting the selective and non-selective operation, respectively, as performed by f, the two effects measured together are $F_1 = F = \phi^*1$ and $F_2 = \tilde{\phi}^*G$, according to (6.1). In this case, the correlation effects (6.19) are

$$F_{12} = \phi^*G \ , \quad F_{12'} = \phi^*G' \ , \quad F_{1'2} = \phi'^*G \ , \quad F_{1'2'} = \phi'^*G' \ . \qquad (6.26)$$

To show this, consider microsystems in a state W. They first trigger the apparatus f - i.e., the effect F_1 - with probability $tr(FW) = tr(\phi W)$. Those systems which have triggered f go into the new state $\hat{W} = \phi W/tr(\phi W)$, thus triggering afterwards the apparatus g - i.e., the effect F_2 - with probability $tr(G\hat{W})$. The probability for the successive triggering of both f and g - i.e., for the occurrence of the effect "F_1 and F_2" = F_{12} - is the product of these two probabilities, $tr(\phi W)\cdot tr(G\hat{W}) = tr(G\cdot\phi W) = tr(\phi^*G\cdot W)$. This implies $F_{12} = \phi^*G$. (We have already presented this argument in §2.) The remaining equations in (6.26) follow similarly. Eqs. (6.20) and (6.21) are satisfied, since

$$F_{12} + F_{12'} = \phi^*(G + G') = \phi^*1 = F = F_1 \ ,$$

$$F_{12} + F_{1'2} = (\phi^* + \phi'^*)G = \tilde{\phi}^*G = F_2 \ ,$$

and

$$F_{12} + F_{12'} + F_{1'2} + F_{1'2'} = \phi^*(G + G') + \phi'^*(G + G')$$

$$= \phi^*1 + \phi'^*1 = F + F' = 1 \ .$$

(Alternatively, we could use Eqs. (6.20) and (6.21) to calculate the last three correlation effects in (6.26) from F_1, F_2 and $F_{12} = \phi^*G$, as remarked above.)

We shall now derive from (6.24) some general results on pairs of coexistent effects.

<u>1.</u> Two effects F_1 and F_2 with $F_1 \leq F_2$ are coexistent.

To show this, take $F_{12'} = 0$. Then, by (6.20), we must set $F_{12} = F_1$ and $F_{1'2} = F_2 - F_1$ (the latter being ≥ 0 since $F_2 \geq F_1$, and $\leq F_2 \leq 1$), and (6.22) is valid since $F_{12} + F_{12'} + F_{1'2} = F_2 \leq 1$. The possibility of choosing $F_{12'}$ = "F_1 and not F_2" = 0 means that, on a suitable apparatus for the joint measurement of F_1 and F_2, the occurrence of F_1 is always accompanied by the occurrence of F_2; i.e., on that apparatus the occurrence of F_1 <u>implies</u> the occurrence of F_2.

<u>2.</u> Two effects F_1 and F_2 with $F_1 \leq F_2'$ are coexistent. (Note that the relation $F_1 \leq F_2'$ is equivalent to $F_1 + F_2 \leq 1$, and is thus symmetric with respect to F_1 and F_2.)

Set $F_{12} = 0$ and thus, by (6.20), $F_{12'} = F_1$, $F_{1'2} = F_2$; then, indeed, $F_{12} + F_{12'} + F_{1'2} = F_1 + F_2 \leq 1$. Clearly, $F_{12} = 0$ means that F_1 and F_2 <u>exclude each other</u> (i.e., never occur together) on an apparatus described by this choice.

<u>3.</u> Two effects F_1 and F_2 with $[F_1,F_2] = 0$ are coexistent.

To show this, set

$$F_{12} = F_1F_2 \ , \quad F_{12'} = F_1F_2' \ , \quad F_{1'2} = F_1'F_2 \ , \quad F_{1'2'} = F_1'F_2' \ . \qquad (6.27)$$

These operators are ≥ 0 and ≤ 1 since, e.g.,

$$(f,F_1F_2f) = \|F_1^{1/2}F_2^{1/2}f\|^2$$

is ≥ 0 and $\leq \|f\|^2 = (f,1f)$ for all $f \in H$. The validity of (6.20) and (6.21) is easily checked. As the four operators (6.27) commute with each other, the same is true for arbitrary sums of them, and thus also for arbitrary effects $F_\alpha \in B$.

The joint measurement of F_1 and F_2 described by (6.27) may be realized as a successive measurement, as in the example (6.1) considered above: First, apply an effect apparatus f_1 performing the complementary operations

$$\phi : W \to F_1^{1/2}\, W\, F_2^{1/2}, \qquad \phi' : W \to F_1'^{1/2}\, W\, F_1'^{1/2}, \qquad (6.28)$$

and thus measuring the effect $\phi^*1 = F_1$; after this, apply an apparatus f_2 measuring F_2. Then, by (6.1) and the fact that F_1 and F_2 commute, the combined apparatus measures together F_1 and

$$\tilde{\phi}^*F_2 = \phi^*F_2 + \phi'^*F_2 = F_1^{1/2}F_2F_1^{1/2} + F_1'^{1/2}F_2F_1'^{1/2}$$

$$= (F_1 + F_1')F_2 = F_2 ,$$

as desired, while (6.26) - with $G = F_2$ and ϕ, ϕ' from (6.28) - immediately leads to (6.27).

The converse of statement 3 is not true, however:

4. The operators F_1 and F_2 describing coexistent effects need not commute.

As an example, consider two projection operators E and $\tilde{E}$ with $[E,\tilde{E}] \neq 0$, and set

$$F_1 = \frac{1}{2} E , \quad F_2 = \frac{1}{2} E + \frac{1}{2} \tilde{E} .$$

(Since $(f,F_2f) = (f,Ef)/2 + (f,\tilde{E}f)/2 \leq \|f\|^2$ for all $f \in H$, F_2 is also an effect, i.e., ≤ 1.) Then we have

$$[F_1,F_2] = \frac{1}{4} [E,\tilde{E}] \neq 0 ;$$

but as $F_1 \leq F_2$, F_1 and F_2 are coexistent by 1.

Commutativity is necessary for coexistence, however, if at least one of the two effects considered is a decision effect:

5. A decision effect E_1 and an arbitrary effect F_2 are coexistent if and only if $[E_1,F_2] = 0$. In this case, all operators F_α in the coexistent completion B of $\{E_1,F_2\}$ are unique and mutually commuting; in particular, the correlation effects are

$$F_{12} = E_1F_2 , \quad F_{12'} = E_1F_2' , \quad F_{1'2} = E_1'F_2 , \quad F_{1'2'} = E_1'F_2' . \qquad (6.29)$$

To show this, we first prove a mathematical statement:

Lemma: Let E be a projection operator and F an arbitrary operator satisfying $0 \leq F \leq E$; then $EF = FE = F$.

Proof: Consider an arbitrary vector $f \in H$ with $Ef = 0$. Then we get $\|F^{1/2}f\|^2 = (f,Ff) \leq (f,Ef) = 0$, i.e., $F^{1/2}f = 0$, and thus also $Ff = 0$. Since $E(E'g) = 0$ for all $g \in H$, we get from this $F(E'g) = 0$, i.e., $FE' = F(1 - E) = 0$, which implies $FE = F = F^* = (FE)^* = EF$.

A part of the statement 5 already follows from 3: If $[E_1,F_2] = 0$, then E_1 and F_2 are coexistent, and (6.29) represents a possible choice of the correlation effects (cf. (6.27)). Now assume E_1 and F_2 to be coexistent. Then, by (6.20), $E_1 = F_{12} + F_{12'}$, and thus $F_{12} \leq E_1$, so that, by the lemma,

$$E_1F_{12} = F_{12}E_1 = F_{12} . \tag{6.30}$$

Similarly, $E_1' = F_{1'2} + F_{1'2'}$ (cf. (6.25)) leads to

$$E_1'F_{1'2} = F_{1'2}E_1' = F_{1'2} ,$$

from which, with $E_1E_1' = E_1'E_1 = 0$, we get

$$E_1F_{1'2} = F_{1'2}E_1 = 0 . \tag{6.31}$$

Eqs. (6.30) and (6.31) show that E_1 commutes with $F_2 = F_{12} + F_{1'2}$, and imply $E_1F_2 = F_{12}$, the first of Eqs. (6.29). The other three equations then follow from (6.20) and (6.21). With the four effects (6.29), all other effects $F_\alpha \in B$ are also uniquely determined. Their mutual commutativity follows as in 3.

As a particular case of 5, we obtain the well-known commensurability criterion of "ordinary" quantum mechanics:

6. Two decision effects E_1 and E_2 are coexistent if and only if $[E_1,E_2] = 0$. In this case, B consists of mutually commuting decision effects (projection operators), $F_\alpha = E_\alpha$, which are uniquely determined by E_1 and E_2. With the operations $\wedge$, $\vee$ and ' defined for commuting projection operators by

$$E_\alpha \wedge E_\beta = E_\alpha E_\beta \ , \quad E_\alpha \vee E_\beta = E_\alpha + E_\beta - E_\alpha E_\beta \ , \quad E'_\alpha = 1 - E_\alpha \ , \tag{6.32}$$

B is a Boolean algebra, and the mapping $\mathbf{F} : \underline{B} \to B$ is a homomorphism, i.e.,

$$E_{\alpha\wedge\beta} = E_\alpha \wedge E_\beta \ , \quad E_{\alpha\vee\beta} = E_\alpha \vee E_\beta \ , \quad E_{\alpha'} = E'_\alpha \ . \tag{6.33}$$

The first statement and the uniqueness of all $F_\alpha \in B$ follow from 5. In the particular case considered, the correlation effects (6.29) are described by projection operators,

$$F_{12} = E_1E_2 = E_{12} \ , \quad F_{12'} = E_1E'_2 = E_{12'} \ , \ \text{etc.} \tag{6.34}$$

They are not only mutually commuting but even mutually orthogonal; i.e., $E_{12}E_{12'} = 0$, etc., which, as is well known, means that they project onto mutually orthogonal subspaces of H. Furthermore, Eq. (6.21),

$$E_{12} + E_{12'} + E_{1'2} + E_{1'2'} = 1 \ ,$$

means that H is the direct sum of the four subspaces $E_{12}H$ to $E_{1'2'}H$. Since all operators $E_\alpha \in B$ are sums of $n \leq 4$ different projection operators from (6.34), they likewise commute among themselves, and are also projection operators E_α – each E_α projects onto the direct sum of the ranges of the n projection operators from (6.34) which add up to E_α.

Consider now two such operators, E_α and E_β. When α and β are represented as disjoint unions of elements of $\underline{B}$ from (6.6), E_α and E_β

become analogous sums of the corresponding operators from (6.34). As can easily be seen, $\alpha \wedge \beta$ is the disjoint union of those elements from (6.6) which occur in both α and β, whereas $\alpha \vee \beta$ is the disjoint union of those elements from (6.6) which occur in at least one of the representations of α and β. On the other hand, since $E_{12}E_{12'} = 0$, etc., $E_\alpha E_\beta$ is the sum of those operators from (6.34) which occur in both sums representing E_α and E_β, and thus coincides with $E_{\alpha \wedge \beta}$, whereas $E_\alpha + E_\beta - E_\alpha E_\beta$ is the sum of those operators from (6.34) which occur in E_α, E_β or both of them, thus being equal to $E_{\alpha \vee \beta}$. ($E_\alpha E_\beta$ has to be subtracted from $E_\alpha + E_\beta$ in order to avoid doublecounting of the operators (6.34) which occur in both E_α and E_β. Compare also Eq. (6.14).) With $\wedge$, $\vee$ and $'$ as defined by (6.32) for projection operators, we have thus proved the first two equations in (6.33), whereas the last one already follows from (6.13). The preceding argument also yields the known geometric interpretation of the operations $\wedge$ and $\vee$ defined by (6.32): $E_\alpha \wedge E_\beta = E_\alpha E_\beta$ projects onto the intersection $E_\alpha H \cap E_\beta H$ of the two subspaces $E_\alpha H$ and $E_\beta H$, whereas $E_\alpha \vee E_\beta = E_\alpha + E_\beta - E_\alpha E_\beta$ projects onto the subspace $E_\alpha H + E_\beta H$ spanned by $E_\alpha H$ and $E_\beta H$. As is also well known, $E'_\alpha = 1 - E_\alpha$ projects onto the orthogonal complement of the subspace $E_\alpha H$.

The Boolean algebra $\underline{B}$ is closed with respect to the operations $\wedge$, $\vee$ and $'$. By (6.33), then, the set of projection operators B is closed with respect to the analogous operations defined by (6.32). When applied to mutually commuting projection operators, these operations (6.32) are well known - and easily checked - to satisfy the calculational rules (6.3) and (6.4) of a Boolean algebra, with the rôles of $\emptyset$ and I taken by the operators 0 and 1, respectively; e.g.,

$$(E_\alpha \wedge E_\beta)' = 1 - E_\alpha E_\beta = (1 - E_\alpha) + (1 - E_\beta) - (1 - E_\alpha)(1 - E_\beta)$$
$$= E'_\alpha + E'_\beta - E'_\alpha E'_\beta = E'_\alpha \vee E'_\beta \; .$$

By (6.10) and (6.33), the mapping $\mathbf{F} : \underline{B} \to B$ preserves the Boolean algebra structure; i.e., it is a homomorphism. As $\alpha \neq \beta$ does not necessarily imply $E_\alpha \neq E_\beta$, the mapping $\mathbf{F}$ is not in general one-to-one (i.e., an isomorphism).

It is also well known that the operations $\wedge$, $\vee$ and $'$ defined by (6.32) for coexistent decision effects may be interpreted as "and", "or" and "not", respectively. This now follows immediately from (6.33) and the physical meaning of the effects $E_{\alpha\wedge\beta}$, $E_{\alpha\vee\beta}$ and $E_{\alpha'}$.

7. In general, the effects F_α, $\alpha \in \underline{B}$ are not uniquely determined by the two coexistent effects F_1 and F_2; i.e., a given coexistent set of effects $\{F_1,F_2\}$ may have several different coexistent completions B.

According to 5, such non-uniqueness is possible only if neither F_1 nor F_2 is a decision effect. As a very instructive example [1], consider the effects

$$F_1 = \frac{1}{2} E \; , \quad F_2 = \frac{1}{2} E + E' \tag{6.35}$$

with a "nontrivial" projection operator E (i.e, $E \neq 0$ or 1).

Since $F_2 = F_1'$, F_1 and F_2 are coexistent, and we may take

$$\left.\begin{aligned} &F_{12} = 0 \; , \quad \text{and thus } F_{12'} = F_1 = \frac{1}{2} E \; , \\ &F_{1'2} = F_2 = \frac{1}{2} E + E' \; , \quad F_{1'2'} = 0 \; , \end{aligned}\right\} \tag{6.36}$$

according to 2. On the other hand, since $F_1 \leq F_2$ as well, we may also set

$$\left.\begin{aligned} &F_{12'} = 0 \; , \quad \text{and thus } F_{12} = F_1 = \frac{1}{2} E \; , \\ &F_{1'2} = F_2 - F_1 = E' \; , \quad F_{1'2'} = \frac{1}{2} E \; , \end{aligned}\right\} \tag{6.37}$$

as in 1. Finally, since $[F_1,F_2] = 0$, a third possibility is to take

$$\left.\begin{aligned} F_{12} &= F_1F_2 = \frac{1}{4}E , \quad F_{12'} = F_1F_2' = \frac{1}{4}E , \\ F_{1'2} &= F_1'F_2 = F_2^2 = \frac{1}{4}E + E' , \quad F_{1'2'} = F_1'F_2' = F_2F_1 = \frac{1}{4}E , \end{aligned}\right\} \quad (6.38)$$

as in 3. For the example (6.35), therefore, the coexistence condition (6.16) may be satisfied with (at least) three different mappings $\mathbf{F} : \alpha \in \underline{B} \to F_\alpha \in L(H)$. These three mappings differ from each other not only in yielding different effects F_α for suitable $\alpha \in \underline{B}$ (e.g., for $\alpha = 1 \wedge 2$), but their ranges $B = \mathbf{F}\underline{B} \subset L(H)$ are also different. Indeed, as can easily be proved, the choice (6.36) yields the set

$$B = \left\{0, \frac{1}{2}E, \frac{1}{2}E + E', 1\right\} ,$$

wheras (6.37) and (6.38) lead to

$$B = \left\{0, \frac{1}{2}E, E, E', \frac{1}{2}E + E', 1\right\}$$

and

$$B = \left\{0, \frac{1}{4}E, \frac{1}{2}E, \frac{3}{4}E, \frac{1}{4}E + E', \frac{1}{2}E + E', \frac{3}{4}E + E', 1\right\} ,$$

respectively.

As shown before, the mapping $\mathbf{F}$ is completely determined by specifying the correlation effects (6.19), and it even suffices to specify one of them if, as in the case considered here, the effects F_1 and F_2 are also given. If, therefore, $\mathbf{F}$ is not uniquely determined by F_1 and F_2, this simply means that the statistical correlations between the results of joint measurements of F_1 and F_2 on single microsystems depend not only on these effects themselves, but also on the particular apparatus used for measuring them together. (Note that, according to our interpretation of the coexistence condition (6.16), every possible choice of $\mathbf{F}$ should be realizable by means of a suitable measuring apparatus.) In view of the fact that effects F represent equivalence classes rather than particular effect apparatuses f, such possible apparatus dependence of the correlations between coexistent effects should not be too surprising; on the contrary, one might rather be surprised that in particular cases, as specified above in 5 and 6, these correlations turn out to be independent of the choice of the measuring apparatus.

For our example (6.35), suitable apparatuses realizing the three different mappings **F** listed above can be visualized quite easily. The choice (6.36) corresponds to the obvious and repeatedly mentioned possibility of measuring together the two complementary effects F_1 and $F_2 = F_1'$: take an arbitrary effect apparatus f_1 measuring F_1, and define the occurrence of F_2 as the non-occurrence of F_1 at this apparatus. Then, by definition, F_1 and F_2 exclude each other, i.e., $F_{12} = 0$, as in (6.36). To realize the choice (6.37), consider an apparatus for the joint measurement of $F_1 = E/2$ and the decision effect $E_2 = E'$. These effect are coexistent, by 5, and they exclude each other irrespective of the chosen apparatus since "F_1 and E_2" $= F_1E_2 = EE'/2 = 0$, by (6.29). This apparatus also measures the effect "F_1 or E_2" which - as "F_1 and E_2" $= 0$ - actually means "either F_1 or E_2", thus coinciding (cf. Eq. (6.15)) with $F_1 + E_2 = E/2 + E' = F_2$. Moreover, by definition, the occurrence of F_1 implies the occurrence of "F_1 or E_2" $= F_2$ at this apparatus, in accordance with the choice $F_{12'} = 0$ in (6.37). (Compare 1.) Finally, as a particular case of (6.27), the choice (6.38) may be realized by the successive application of suitable apparatuses f_1 and f_2 measuring F_1 and F_2, respectively, as explained in 3.

Some remarkable consequences of the possible apparatus dependence of the correlaions between coexistent effects may also be illustrated by the example (6.35).

According to 2, two effects F_1 and F_2 with $F_1 \leq F_2'$ exclude each other on a suitable measuring apparatus. This need not be so, however, if another apparatus is used. As an example, consider the two complementary effects F_1 and F_2 given by (6.35), as measured together by an apparatus described by (6.37): In this case, the occurrence of F_1 does not exclude, but rather implies the occurrence of F_2. (In view of such possibilities, it might appear a little misleading to denote the effect F' by "not F". However, this expresses quite suggestively the obvious and most "natural" possibility of measuring F' together with F, as realized in the example (6.35) by the choice (6.36).)

If two effects F_1 and F_2 satisfy $F_1 \leq F_2$, then on a suitable apparatus the occurrence of F_1 implies the occurrence of F_2, according to 1. Again this need not be true for other apparatuses. To exemplify this, take again the effects F_1 and F_2 from (6.35): On an apparatus described by (6.36), the occurrence of F_1 excludes the occurrence of F_2, rather than implying it.

Two identical effects, $F_1 = F_2 = F$, are always coexistent. According to 1, we may choose $F_{12'} = 0$, which implies $F_{12} = F$ and $F_{1'2} = 0$ in this case. On an apparatus described by this choice, F_1 implies F_2 (as $F_{12'} = 0$), and vice versa (as $F_{1'2} = 0$); i.e., F_1 and F_2 always occur together. Such an apparatus can be realized simply by feeding the output of an apparatus f measuring F into two different output channels. There are also less trivial possibilities, however, of measuring the same effect F by reading two different output channels of one apparatus: Consider again the effects (6.35) and an apparatus corresponding to (6.37) for their joint measurement. Then the two correlation effects F_{12} and $F_{1'2'}$ both coincide with the effect $F = E/2$, but rather than always occuring together, they actually exclude each other.

Such things can not happen, however, if at least one of the two effects F_1 and F_2 is a decision effect, say $F_1 = E_1$. Then, if $E_1 \leq F_2'$, E_1 and F_2 always exclude each other: $E_1 \leq F_2'$ is equivalent to $F_2 \leq E_1'$, therefore the lemma in 5 implies $E_1'F_2 = (1 - E_1)F_2 = F_2$, so that $F_{12} = E_1F_2 = 0$, by (6.29). In particular, complementary decision effects E and E' always exclude each other. Moreover, F_2 implies E_1 if $F_2 \leq E_1$, and E_1 implies F_2 if $E_1 \leq F_2$. In the first case, the lemma in 5 yields $E_1F_2 = F_2$ so that, by (6.29), $F_{1'2} = E_1'F_2 = 0$; i.e., F_2 implies E_1. In the second case, $E_1 \leq F_2$ yields $F_2' \leq E_1'$, so that, again by the lemma, $E_1'F_2' = F_2'$ and thus, by (6.29), $F_{12'} = E_1F_2' = 0$; i.e., E_1 implies F_2. Finally, if some apparatus performs together, when applied to a single microsystem, two or more measurements of one and the same decision effect E, then the results of all these measurements must be identical in every single experiment. Indeed, if two outputs 1 and 2 of some apparatus both correspond to measurements of E, i.e., $F_1 = F_2 = E$, then (6.29) implies $F_{12'} = F_{1'2} = EE' = 0$ which, as already explained, yields the desired conclusion.

The last-mentioned property is characteristic for decision effects. I.e., if F is not a decision effect ($F^2 \neq F$), then there exists an apparatus which performs together two F measurements on single microsystems in such a way, that at least sometimes the results of these two measurements are different from each other. To show this, replace E by F in the preceding argument, and choose the correlation effects according to (6.27); this now yields $F_{12'} = F_{1'2} = F - F^2 \neq 0$, so that the probabilities for obtaining "yes" in the first and "no" in the second measurement, as given by $tr(F_{12'}W)$, are non-zero for suitable states W. We may express this result more succinctly: Different measurements of the same effect F on the same microsystem need not always give identical results unless F is a decision effect. In this respect, therefore, the results obtained by measuring decision effects E turn out to be less apparatus dependent, and thus appear more like "properties" of the microsystem itself, than the results of measurements of other effects F.

According to 5, it is sufficient for the uniqueness of the mapping $\mathbf{F} : \alpha \to F_\alpha$ that at least one of the two effects F_1 and F_2 is a decision effect. This condition is not necessary, however, as shown by the following example. Take $F_1 = \alpha E$ and $F_2 = \beta E'$ with a (nontrivial) projection operator E and real numbers α and β satisfying $0 < \alpha \leq \beta < 1$. Since $[F_1, F_2] = 0$, F_1 and F_2 are coexistent. Condition (6.22) may be replaced here, using (6.20) and $F_{12} \geq 0$, by the stronger estimate

$$F_{12} + F_{12'} + F_{1'2} \leq F_1 + F_2 = \alpha E + \beta E' \leq \beta(E + E') = \beta 1 \ . \qquad (6.39)$$

The effects $G_1 = F_1/\beta = \alpha E/\beta$ and $G_2 = F_2/\beta = E'$ are also coexistent, and by (6.20) they may be written in the form

$$G_1 = G_{12} + G_{12'} \ , \qquad G_2 = G_{12} + G_{1'2}$$

with correlation effects $G_{12} = F_{12}/\beta$, etc., which indeed satisfy

$$G_{12} + G_{12'} + G_{1'2} \leq 1 \ ,$$

according to (6.39). But because $G_2 = E'$ is a decision effect, the correlation effect G_{12} is unique (cf. 6). Therefore the original correlation effect $F_{12} = \beta G_{12}$ is also unique, which implies uniqueness of $\mathbf{F}$.

8. Consider, as in §4, a composite system consisting of two noninteracting subsystems I and II. Then arbitrary pairs of subsystem effects (cf. (4.10)),

$$\underline{F}_1 = F_I \otimes 1_{II} \ , \qquad \underline{F}_2 = 1_I \otimes F_{II} \ , \qquad (6.40)$$

are coexistent, according to 3, and a possible choice for the correlation effect $\underline{F}_{12}$ = "$\underline{F}_1$ and $\underline{F}_2$" is, by (6.27),

$$\underline{F}_{12} = \underline{F}_1\underline{F}_2 = F_I \otimes F_{II} \ . \qquad (6.41)$$

If F_I or F_{II} (and thus $\underline{F}_1$ or $\underline{F}_2$) is a decision effect, then (6.41) is the only possible choice, thus leading to a unique mapping $\mathbf{F} : \underline{B} \to B \subset L(\underline{H})$. Moreover, as discussed in detail in §4, the choice (6.41) is in a certain sense the "natural" one also if neither F_I nor F_{II} is a decision effect, since it corresponds to the simplest and therefore most "natural" possibility for the joint measurement of the subsystem effects $\underline{F}_1$ and $\underline{F}_2$. (See the derivation of (6.41) - which there occurs as Eq. (4.9) - in §4.)

Nevertheless, there are also pairs of subsystem effects (6.40) for which (6.41) does not represent the only possible choice. As an example, set

$$\underline{F}_1 = \frac{1}{2} E_I \otimes E_{II} , \quad \underline{F}_2 = 1_I \otimes \frac{1}{2} E_{II}$$

with two nontrivial projection operators E_I and E_{II}. In this case, we may take

$$\underline{F}_{12} = \alpha(E_I \otimes E_{II})$$

with an arbitary real α between 0 and 1/2. Eqs. (6.20) then yield $\underline{F}_{12'} = \underline{F}_1 - \underline{F}_{12}$ and $\underline{F}_{1'2} = \underline{F}_2 - \underline{F}_{12}$; these two operators really belong to $L(\underline{H})$ since, e.g.,

$$\underline{1} \geq \underline{F}_1 \geq \underline{F}_1 - \underline{F}_{12} = \underline{F}_{12'} = \frac{1}{2}(E_I \otimes 1_{II}) - \alpha(E_I \otimes E_{II})$$

$$\geq \frac{1}{2}((E_I \otimes 1_{II}) - (E_I \otimes E_{II})) = \frac{1}{2}(E_I \otimes E'_{II}) \geq 0 .$$

Moreover, since both $\underline{F}_1$ and $\underline{F}_2$ are $\leq \underline{1}/2$, we have

$$\underline{F}_{12} + \underline{F}_{12'} + \underline{F}_{1'2} = \underline{F}_1 + \underline{F}_2 - \underline{F}_{12} \leq \underline{F}_1 + \underline{F}_2 \leq \underline{1} ,$$

so that (6.22) is also satisfied.

Besides illustrating the possible non-uniqueness of the correlations between subsystem effects, the above example also shows that there may exist infinitely many different mappings $\mathbf{F} : \underline{B} \rightarrow B$ for a given pair of coexistent effects, as parametrized here by the number α. Except for the particular case $\alpha = 1/4$ - which corresponds to the "natural" choice (6.41) for $\underline{F}_{12}$ - it is completely unknown, however, how apparatuses realizing all these mappings $\mathbf{F}$ would look in practice.

We conclude our discussion of the particular case of two coexistent effects F_1 and F_2 with the construction of a quantum mechanical model for their joint measurement. As in §5, the model apparatus is described as a quantum mechanical system with state space H_a and initial state W_a, and is assumed to form with the microsystem considered a binary scattering system, characterized by a unitary scattering operator $\underline{S}$ on $\underline{H} = H \otimes H_a$. But now the apparatus is assumed to have two output channels, rather than only a single one. Their yes-no outputs have to be "read" at the apparatus after its interaction with the microsystem. This "reading" thus consists of two yes-no measurements performed together, as described by a coexistent pair of effect operators E_1^a and E_2^a on H_a. Actually the "reading" is performed on the composite system, and is therefore described by the two effects $1 \otimes E_1^a$ and $1 \otimes E_2^a$ on $\underline{H}$. As is obvious, e.g., from the condition (6.24), the latter are indeed coexistent if E_1^a and E_2^a are. (The converse is equally obvious - e.g., from $\underline{6}$ - if E_1^a and E_2^a are decision effects. Although, as our notation already suggests, we will assume this later on, at the moment E_1^a and E_2^a may still be arbitrary effects.)

A model apparatus of this kind measures together two effects F_1 and F_2 of the microsystem, defined implicitly but uniquely by the equations

$$\mathrm{tr}(F_i W) = \mathrm{tr}(1 \otimes E_i^a)\underline{S}(W \otimes W_a)\underline{S}^*), \quad i = 1 \text{ or } 2, \tag{6.42}$$

with $W \in K(H)$ arbitrary. (Compare Eq. (5.1), and remember that the right hand side of (6.42) is the probability for output i to be "yes" after the interaction of the apparatus with microsystems in the state W.) As expected, F_1 and F_2 are coexistent. In order to prove this, we represent the coexistent "output" effects E_1^a and E_2^a, in analogy to (6.20), in the form

$$E_1^a = E_{12}^a + E_{12'}^a, \quad E_2^a = E_{12}^a + E_{1'2}^a, \tag{6.43}$$

with four effects E_{ij}^a (i = 1 or 1', j = 2 or 2') describing the correlations between the two "output" effects E_1^a and E_2^a and satisfying, in analogy to (6.21),

$$E^a_{12} + E^a_{12'} + E^a_{1'2} + E^a_{1'2'} = 1_a \,. \tag{6.44}$$

Corresponding effects F_{ij} of the microsystem can then be defined, analogous to (6.42), by

$$tr(F_{ij}W) = tr((1 \otimes E^a_{ij})\underline{S}(W \otimes W_a)\underline{S}^*), \quad i = 1 \text{ or } 1', \; j = 2 \text{ or } 2'. \tag{6.45}$$

These effects F_{ij} have to be interpreted as correlation effects for F_1 and F_2 (cf. (6.19)) since, e.g., with i = 1 and j = 2', the right hand side of (6.45) is the probability for the occurrence of the apparatus effect $E^a_{12'}$ = "E^a_1 and not E^a_2" after the interaction, which - by definition of the effects F_1 and F_2 - is equivalent to the occurrence of the effect "F_1 and not F_2". Eqs. (6.20) and (6.21) for F_1, F_2 and the correlation effects F_{ij} are easily derived from Eqs. (6.43) and (6.44) and the definitions (6.42) and (6.45). (For instance, (6.42), (6.43) and (6.45) lead to

$$tr((F_{12} + F_{12'})W) = tr((1 \otimes [E^a_{12} + E^a_{12'}])\underline{S}^*) = tr(F_1W)$$

for all W, which implies $F_1 = F_{12} + F_{12'}$.)

Now consider, conversely, an arbitrarily given pair $\{F_1, F_2\}$ of coexistent effects, and a representation (6.20) of F_1 and F_2 in terms of given - but, in cases of nonuniqueness, deliberately chosen - correlation effects F_{ij}. We will prove that the joint measurement of F_1, F_2 and the given correlation effects F_{ij} can be "realized" by a model apparatus of the type considered. According to the previous discussion, this amounts to proving the existence of a Hilbert space H_a, a state $W_a \in K(H_a)$, a unitary operator $\underline{S}$ on $H \otimes H_a$, and four effects $E^a_{ij} \in L(H_a)$ (i = 1 or 1', j = 2 or 2') satisfying (6.44), such that Eqs. (6.45) are satisfied with the given correlation effects $F_{ij} \in L(H)$ and arbitrary states $W \in K(H)$. With coexistent "output" effects E^a_1 and E^a_2 defined by (6.43), then, Eqs. (6.45) and (6.20) imply (6.42), so that the apparatus indeed measures F_1 and F_2, with correlations described by F_{12}, $F_{12'}$, etc. As remarked previously, this apparatus then also measures - at least implicitly - all other effects $F_\alpha \in B$.

We shall assume that, as already suggested by the notation, the E^a_{ij} are decision effects. Then (6.44) means that the corresponding projection operators project onto four mutually orthogonal subspaces $E^a_{ij}H_a$, and H_a is the direct sum of the latter. Moreover, (6.43) defines two commuting projection operators E^a_1 and E^a_2, and we get

$$E^a_{12} = E^a_1 E^a_2 \ , \qquad E^a_{12'} = E^a_1 E^a_{2'} \ , \qquad \text{etc.},$$

in accordance with our previous results (see 6) and with "conventional" quantum mechanics. Choosing four operations ϕ_{ij} (i = 1 or 1', j = 2 or 2') with

$$\phi^*_{ij} 1 = F_{ij} \tag{6.46}$$

but arbitrary otherwise, we replace (6.45) by the stronger requirements

$$\phi_{ij} W = \mathrm{Tr}_a((1 \otimes E^a_{ij}) \underline{S}(W \otimes W_a) \underline{S}^*) \ . \tag{6.47}$$

(By taking the trace, (6.47) is easily seen to imply (6.45), by virtue of (6.46). Being analogous to (5.14), (6.47) means that the operation ϕ_{ij} is performed by selecting those microsystems which have triggered the effect E^a_{ij} at the apparatus - i.e., the correlation effect F_{ij}.)

We are thus left with the problem of representing four given operations ϕ_{ij} satisfying, by (6.46) and (6.21),

$$(\phi^*_{12} + \phi^*_{12'} + \phi^*_{1'2} + \phi^*_{1'2'}) 1 = 1 \ , \tag{6.48}$$

in the form (6.47), with suitable W_a, $\underline{S}$, and four projection operators E^a_{ij} satisfying (6.44). An analogous problem has already been solved in §5 in the proof of Theorem 2. There two given operations ϕ and ϕ' satisfying

$$(\phi^* + \phi'^*)1 = 1 ,$$

in analogy to (6.48), were represented in a form analogous to (6.47) (cf. (5.20) and (5.21)),

$$\phi^{(')}W = Tr_a((1 \otimes E_a^{(')})\underline{S}(W \otimes W_a)\underline{S}^*) ,$$

with two projection operators E_a and E'_a which satisfy the condition

$$E_a + E'_a = 1_a$$

analogous to (6.44). As can be easily seen by inspection, the explicit construction of H_a, W_a, $\underline{S}$, E_a and E'_a described in §5 can be generalized immediately to the present problem. This establishes the existence of the desired quantum mechanical model.

The model apparatus can also be "used" to perform operations; e.g., as already remarked, the operations ϕ_{ij}, by selecting the microsystems which have triggered the correlation effects F_{ij}; or the operations

$$\phi_1 = \phi_{12} + \phi_{12'} \quad \text{and} \quad \phi_2 = \phi_{12} + \phi_{1'2} ,$$

by selecting the microsystems which have triggered the effects F_1 or F_2, respectively; or the non-selective operation

$$\tilde{\phi} = \phi_{12} + \phi_{12'} + \phi_{1'2} + \phi_{1'2'} .$$

For a given pair of coexistent effects F_1, F_2 and given correlation effects F_{ij}, the choice of the four operations ϕ_{ij} (cf. (6.46)) is still highly arbitrary. The same, therefore, is true also for operations like ϕ_1, ϕ_2 and $\tilde{\phi}$.

The coexistence criterion (6.16) may be generalized immediately to arbitrary sets of effects, as follows:

$$\left.\begin{array}{l}\text{A set } C \subset L(H) \text{ is coexistent if and only if there} \\ \text{exist a Boolean algebra } \underline{B} \text{ and a mapping} \\ \qquad\qquad \mathbf{F} : \alpha \in \underline{B} \rightarrow F_\alpha \in L(H) \\ \text{satisfying (6.10) and (6.12), such that } C \text{ is con-} \\ \text{tained in the range } B = \mathbf{F}\underline{B} = \{F_\alpha \mid \alpha \in \underline{B}\} \text{ of } \mathbf{F}.\end{array}\right\} \quad (6.49)$$

Namely, if C is coexistent, there exists an apparatus c for the joint measurement of all effects $F \in C$. As described above for the particular case $C = \{F_1, F_2\}$, this apparatus c can be extended by adding new output channels, until one finally arrives at an apparatus b whose outputs α form a Boolean algebra $\underline{B}$, and which measures together all effects $F_\alpha \in \mathbf{F}\underline{B} \supseteq C$. Therefore (6.49) is necessary for coexistence. Conversely, if (6.49) is satisfied, we may consistently assume the existence of such an apparatus b, thus taking (6.49) also as a sufficient condition.

If taken literally, these arguments seem to apply only if the set C and the Boolean algebra $\underline{B}$ both consist of finitely many elements, since apparently a real measuring apparatus c of the type considered here can have finitely many output channels only, and the successive application of the operations $\wedge$, $\vee$ and ' to the finitely many outputs of this apparatus c then leads - as in the case of two outputs discussed before - also to a finite Boolean "output" algebra $\underline{B}$. As we shall see below, however, there are also simple and physically interesting possibilities of measuring together infinite sets of effects satisfying the coexistence criterion (6.49).

With C, obviously, the set of effects $B = \mathbf{F}\underline{B} \supseteq C$ also satisfies the coexistence condition (6.49). Thus B is also coexistent, and is called here - as in the particular case $C = \{F_1, F_2\}$ - a coexistent completion of C.

An arbitrary pair of effects $\{F_1,F_2\}$ chosen from a coexistent set C satisfies condition (6.24), and is therefore coexistent. Indeed, since $\{F_1,F_2\} \subseteq C \subseteq \mathbf{F}\underline{B}$, by (6.49), there are two elements of $\underline{B}$ - 1 and 2, say - which are mapped by $\mathbf{F}$ into F_1 and F_2, respectively. As a Boolean algebra, $\underline{B}$ contains along with 1 and 2 also the four elements (6.6) satisfying Eqs. (6.17) and (6.18). With effects F_{12}, $F_{12'}$ etc. defined by (6.19), then, we obtain Eqs. (6.20) and (6.21) - i.e., condition (6.24) - from (6.17), (6.18) and the properties (6.10) and (6.12) of the mapping $\mathbf{F}$. This argument also shows that (6.49) implies (6.24) for the particular case $C = \{F_1,F_2\}$. Since, on the other hand, (6.16) trivially implies (6.49) in this case, the coexistence condition (6.49) is really a generalization of the two equivalent coexistence conditions (6.16) and (6.24) for pairs of effects.

We do not know whether, conversely, a set C of effects is coexistent, provided this is true for all pairs $\{F_1,F_2\}$ from C. But if C consists of decision effects only, this can indeed be proved. Or, in other words: A set C of decision effects is coexistent if and only if every pair $\{E_1,E_2\}$ from C is coexistent - i.e., according to $\underline{6}$: if and only if C consists of pairwise commuting projection operators.

The "only if" part of this statement has already been proved. Consider, therefore, an arbitrary set C of mutually commuting projection operators. By successively applying to these operators the operations $\wedge$, $\vee$ and ' as defined by (6.32), we obtain from C a set $B \supseteq C$, also consisting of mutually commuting projection operators, which is closed - i.e., a Boolean algebra - with respect to these operations (6.32). In other words, B is the unique Boolean algebra of projection operators generated by C. (We leave aside here some topological questions, which would arise if B were assumed to contain also certain limit elements like, e.g., $\lim_{n\to\infty} E_1 \wedge \ldots \wedge E_n$ with $E_1, E_2 \ldots \in B$.) We introduce again the notation $E_1 \dot{\vee} \ldots \dot{\vee} E_n$ for $E_1 \vee \ldots \vee E_n$ if $E_1 \ldots E_n \in B$ and $E_i \wedge E_j \equiv E_i E_j = 0$ for $i \neq j$. For such E_i, an obvious generalization of (6.4) yields

$$E_1 \wedge (E_2 \vee \ldots \vee E_n) = (E_1 \wedge E_2) \vee \ldots \vee (E_1 \wedge E_n) = 0 ,$$

so that, by (6.32),

$$E_1 \vee (E_2 \vee \ldots \vee E_n) = E_1 + (E_2 \vee \ldots \vee E_n) .$$

Repeating this argument, we thus obtain

$$E_1 \dot{\vee} \ldots \dot{\vee} E_n = E_1 + \ldots + E_n . \tag{6.50}$$

In order to prove that C satisfies the coexistence condition (6.49), we now identify the (abstract) Boolean algebra $\underline{B}$ occuring in (6.49) with the above Boolean algebra B of projection operators generated by C, and take for $\mathbf{F}$ the identity mapping. Then $\mathbf{F}\underline{B} \equiv B$ contains C, by definition. Moreover, since in B the abstract elements I and $\emptyset$ are represented by the operators 1 and 0, respectively, condition (6.10) is trivially satisfied. Finally, for the identity mapping $\mathbf{F}$ the additivity property (6.12) is identical with (6.50). Thus, indeed, the set C is coexistent.

Consider, more generally, a coexistent set C of effects containing a subset C_o of decision effects, a Boolean algebra $\underline{B}$ and a mapping $\mathbf{F}$ as in (6.49). Since $\mathbf{F}\underline{B} = B$ contains C_o, there exists a nonempty subset $\underline{B}'$ of $\underline{B}$ consisting of those elements α which are mapped into decision effects, $F_\alpha = E_\alpha$. With arbitrary α and $\beta \in \underline{B}'$, the two effects E_α and E_β are coexistent - i.e., commuting projection operators. Since $\underline{B}$ is a Boolean algebra, the set $B = \mathbf{F}\underline{B}$ contains also the effects $F_{\alpha \wedge \beta}$ = "E_α and E_β", $F_{\alpha \vee \beta}$ = "E_α or E_β" and $F_{\alpha'}$ = "not E_α". According to $\underline{6}$, the latter are again decision effects, and are given by

$$\left.\begin{aligned} E_{\alpha \wedge \beta} &= E_\alpha E_\beta = E_\alpha \wedge E_\beta , \\ E_{\alpha \vee \beta} &= E_\alpha + E_\beta - E_\alpha E_\beta = E_\alpha \vee E_\beta , \\ E_{\alpha'} &= 1 - E_\alpha = E'_\alpha \end{aligned}\right\} \tag{6.51}$$

(cf. (6.32) and (6.33)). Therefore $\underline{B}'$ contains along with α and β

also $\alpha \wedge \beta$, $\alpha \vee \beta$ and α'; i.e., $\underline{B}'$ is a Boolean subalgebra of $\underline{B}$. Moreover, by (6.51), the image $B' = \mathbf{F}\underline{B}'$ of $\underline{B}'$ under $\mathbf{F}$ - consisting of coexistent decision effects, i.e., commuting projection operators - is closed, and is thus a Boolean algebra, with respect to the operations $\wedge$, $\vee$ and $'$ defined by (6.32), and the restriction of the mapping $\mathbf{F}$ to $\underline{B}'$ is a homomorphism of $\underline{B}'$ onto B'. As in the particular case $\underline{6}$ discussed above, this homomorphism leads to the familiar physical interpretation of the operations $\wedge$, $\vee$ and $'$ for arbitrary projection operators in B'.

As a Boolean algebra, B' contains along with C_o the whole Boolean algebra B_o of projection operators generated by C_o. If thus, in particular, C contains only decision effects (i.e., $C = C_o$), then every coexistent completion B of C contains the Boolean algebra B_o of projection operators generated by C. On the other hand, as shown before, B_o itself is a possible choice for B; hence B_o is the minimal coexistent completion of a given coexistent set C of decision effects.

The concept of an <u>observable</u> is taken as fundamental in some formulations of quantum mechanics. In the approach presented here, this notion appears on the contrary as a derived one. We shall conclude by presenting a short discussion of observables, especially since coexistent sets of effects play an important rôle in this connection.

In accordance with the general ideas discussed in §1, observables are defined here operationally in terms of the apparatuses measuring them. Like a preparing instrument or an effect apparatus, an apparatus measuring an observable is also completely specified by the "classical" description of its construction and application.

An easily visualizable example would be an apparatus with a scale, on which a pointer indicates the "measured value" of the observable after the application of the apparatus to a microsystem. In practice very few - if any - quantum mechanical observables are measured by such simple apparatuses; thus we shall only assume here that somehow each application of the measuring apparatus to a single microsystem yields a well-defined real number as the "measured value". In view of the errors connected with any real measurement, the assumption that these measured values are given with infinite precision is clearly an idealization. We leave aside here the difficult problem of justifying idealizations of this kind, without which a mathematical description of observables would become very cumbersome or even impossible.

By definition, the "scale" of a measuring apparatus contains all possible measured values of the observable considered, but besides them it may also contain real numbers which never occur as measured values. In practice every scale is finite, and could thus be identified with a suitable finite interval on the real line. It is advantageous here, however, to use the whole real line I as a universal scale for all observables. Thereby we also include in the subsequent discussion the case of "unbounded" observables like, e.g., position coordinates or momentum components of a particle, whose possible measured values are not confined to a finite interval, and which therefore represent limiting cases of "real" (i.e., bounded) observables.

Consider now suitable subsets α of the real line I, e.g., intervals; for technical reasons (see below) we choose here the more general class of Borel sets (cf., e.g., [13], Ch. 1). If an observable is measured on a single microsystem, the question wether or not the measured value is contained in a given Borel set α constitutes a yes-no measurement on the given system, and is thus to be described by an effect operator F_α. We thereby obtain, for a given observable, a certain mapping $\mathbf{F} : \alpha \to F_\alpha$ of Borel sets α into L(H). By definition of F_α, $\mathrm{tr}(F_\alpha W)$ represents the probability of finding the measured value in the set α, when the observable considered is measured in the state W. The operators F_α for arbitrary α - or, in other words, the mapping $\mathbf{F}$ - thus completely specify the "statistics" of the given observable in arbitrary states W, and in this sense provide a complete quantum mechanical description of the observable.

In particular, since the measured value will always lie on the real line I and never in the empty set $\emptyset$, we immediately get

$$F_I = 1 \;, \quad F_\emptyset = 0 \;. \tag{6.52}$$

Moreover, clearly, the probability of finding, in a given state W, measured values in the union $\bigcup_i \alpha_i$ of finitely many mutually disjoint sets $\alpha_1 \dots \alpha_n$ (i.e., $\alpha_i \cap \alpha_j = \emptyset$ for $i \neq j$), is the sum of the corresponding probabilities for these sets $\alpha_1 \dots \alpha_n$ separately; i.e.,

$$\mathrm{tr}(F_{\dot{\bigcup}_i \alpha_i} W) = \sum_i \mathrm{tr}(F_{\alpha_i} W) \;. \tag{6.53}$$

Being true for arbitrary W, this implies the additivity property

$$F_{\dot{\bigcup}_i \alpha_i} = \sum_i F_{\alpha_i} \tag{6.54}$$

of the mapping $\mathbf{F} : \alpha \to F_\alpha$. In (6.53) and (6.54) we have written, in analogy to our previous notation, the "disjoint union" of the sets α_i as $\dot{\bigcup}_i \alpha_i$. We have also omitted an upper limit n in $\dot{\bigcup}_i \alpha_i$ and $\sum_i F_{\alpha_i}$, thereby indicating that Eqs. (6.53) and (6.54) are assumed here to be valid not only for finite but also for countably infinite sequences of mutually disjoint Borel sets α_i. This deserves a little explanation. First, a union of countably many (arbitrary, not necessarily disjoint) Borel sets α_i is again a Borel set; therefore the left hand sides of (6.53) and (6.54) exist also for infinitely many sets α_i. Moreover, in this case, the right hand side of (6.54) exists as an ultraweak limit. To show this, consider the finite disjoint unions $\beta_n = \dot{\bigcup}_{i<n} \alpha_i$, for which we already have $F_{\beta_n} = \sum_{i<n} F_{\alpha_i}$, according to the "finite" version of (6.54). The nonnegative operators F_{β_n} thus increase with n. Since, as elements of L(H), they are also bounded from above by 1, they indeed converge ultraweakly for $n \to \infty$ to an operator in L(H), which is taken as defining the right hand side of (6.54). Then, by the definition of ultraweak convergence, the right hand side of (6.53) is also convergent, and (6.54) is equivalent to

(6.53) - with arbitrary $W \in K(H)$ - also for countably infinite disjoint unions. Finally, the physical interpretation of $tr(F_{\dot{\bigcup}_i \alpha_i} W)$ and $tr(F_{\alpha_i} W)$ as probabilities leads to (6.53), regardless of whether $\dot{\bigcup}_i \alpha_i$ is a finite or an infinite disjoint union. (Although in the latter case the argument is not entirely trivial, we shall not present it in detail here.) - Since a set β with $\beta \supseteq \alpha$ may be represented as $\alpha \dot{\cup} \gamma$ with a suitable set γ, (6.54) also implies

$$F_\alpha \leq F_\beta \quad \text{if} \quad \alpha \subseteq \beta \ . \tag{6.55}$$

As is well known, the Borel sets on the real line I form a Boolean algebra $\underline{B}$, if one defines "α and β" = $\alpha \wedge \beta$ as the intersection $\alpha \cap \beta$, "α or β" = $\alpha \vee \beta$ as the union $\alpha \cup \beta$, and "not α" = α' as the complement of the set α. The "trivial" elements of this Boolean algebra (cf., e.g., (6.3)) are the whole real line I and the empty set $\emptyset$, as already indicated by our notation. The fact that $\underline{B}$ contains not only finite but also countably infinite intersections and unions of arbitrary Borel sets is usually expressed by calling $\underline{B}$ a Boolean σ-algebra. The mapping $\mathbf{F} : \underline{B} \to \mathbf{F}\underline{B} = B \subset L(H)$ considered above is a mapping of the kind considered in condition (6.24), since Eqs. (6.52) correspond to (6.10), whereas (6.54) represents a generalization of (6.12). The set of effects $\{F_\alpha \mid \alpha \in \underline{B}\} = B$ is thus a coexistent set.

An observable, therefore, may be described quantum mechanically as a particular coexistent set of effects in the sense of condition (6.24), with $\underline{B}$ standing for the Boolean σ-algebra of Borel sets on the real line I, and with a generalized additivity property (6.54) - which is quite natural for a Boolean σ-algebra - in place of (6.12). This description is also in accordance with the operational meaning of coexistence since, indeed, a measuring apparatus for the given observable measures together all effects $F_\alpha \in \mathbf{F}\underline{B}$, as explained above. Eqs. (6.52) and (6.54) also imply the relations

$$F_{\alpha'} = 1 - F_\alpha = F'_\alpha \tag{6.56}$$

and

$$F_{\alpha \cup \beta} = F_\alpha + F_\beta - F_{\alpha \cap \beta} . \qquad (6.57)$$

(Compare the derivation of (6.13) and (6.14) from (6.10) and (6.12).)

A mapping $\mathbf{F} : \alpha \to F_\alpha$ of Borel sets into effects satisfying the conditions (6.52) and (6.54) is called a positive operator-valued measure - or, more briefly: a POV measure - on the real line. This is motivated by the definition of an "ordinary" (normalized) measure, which as a mapping $\omega : \alpha \to \omega(\alpha)$ of Borel sets α into nonnegative numbers $\omega(\alpha)$ is characterized by the analogous conditions

$$\omega(I) = 1 , \quad \omega(\emptyset) = 0 , \quad \omega(\dot{\bigcup_i} \alpha_i) = \sum_i \omega(\alpha_i) .$$

(In particular, these conditions are satisfied, by virtue of (6.52) and (6.53), for $\omega(\alpha) = \mathrm{tr}(F_\alpha W)$ with a POV measure F_α and an arbitrary state W.)

If only projection operators are considered as describing yes-no measurements, as in "conventional" quantum mechanics, then all effects F_α occuring in the above discussion must be assumed to be decision effects, $F_\alpha = E_\alpha$. Thereby one arrives at a mapping $\mathbf{E} : \alpha \to E_\alpha$ of Borel sets into decision effects (projection operators), which satisfies (6.52) and (6.54) in the form

$$E_I = 1 , \quad E_\emptyset = 0 \qquad (6.58)$$

and

$$E_{\dot{\bigcup_i} \alpha_i} = \sum_i E_{\alpha_i} , \qquad (6.59)$$

and is called a spectral measure on I. In "conventional" quantum mechanics, therefore, every observable corresponds to a spectral measure $\mathbf{E}$ on the real line, whereas in the framework considered here such observables - the so called decision observables - constitute a particular class only. Since the decision effects E_α, $\alpha \in \underline{B}$ form a coexistent set, they are represented by mutually commuting projection operators, and according to (6.51) we also have

$$E_{\alpha \cap \beta} = E_\alpha E_\beta \ . \tag{6.60}$$

(The last-mentioned properties can also be derived directly from (6.59), by using the elementary result that the sum $E + \tilde{E}$ of projection operators E and $\tilde{E}$ is again a projection operator if and only if $E\tilde{E} = \tilde{E}E = 0$. With (6.59), this immediately implies $E_\gamma E_\delta = E_\delta E_\gamma = 0$ for arbitrary disjoint Borel sets γ and δ. For arbitrary sets α and β, Eq. (6.59) and the decompositions $\alpha = (\alpha \cap \beta) \cup (\alpha \cap \beta')$ and $\beta = (\alpha \cap \beta) \cup (\alpha' \cap \beta)$ lead to $E_\alpha = E_{\alpha \cap \beta} + E_{\alpha \cap \beta'}$ and $E_\beta = E_{\alpha \cap \beta} + E_{\alpha' \cap \beta}$. Because the sets $\alpha \cap \beta'$ and $\alpha' \cap \beta$ are disjoint, this finally implies $E_\alpha E_\beta = E_\beta E_\alpha = E_{\alpha \cap \beta}$.) Eq. (6.60) may be trivially generalized to finite intersections $\alpha_1 \cap \dots \cap \alpha_n$, but actually it is true, in the form

$$E_{\bigcap_i \alpha_i} = \prod_i E_{\alpha_i} \ , \tag{6.61}$$

for arbitrary finite or countably infinite intersections of Borel sets α_i. In the latter case, the right hand side of (6.61) exists, and has to be understood, as the ultraweak limit of $\Pi_{i \leq n} E_{\alpha_i}$ for $n \to \infty$. We will not prove this here. Besides (6.61), clearly, Eqs. (6.56) and (6.57) are also true for spectral measures, i.e., with E_α for F_α. (Compare also (6.51).)

The more familiar description of decision observables by <u>self-adjoint operators</u> is obtained as follows. Define a one-parameter family of projection operators $E(\lambda)$ by inserting for α the intervals $(-\infty, \lambda]$ into the spectral measure $\mathbf{E} : \alpha \to E_\alpha$; i.e.,

$$E(\lambda) = E_{(-\infty,\lambda]} \ . \tag{6.62}$$

By (6.55), (6.58) and (6.59), then, $E(\lambda)$ may be shown to have the characteristic properties of a <u>spectral family</u> of a self-adjoint operator A, namely:

i) $\lambda_1 < \lambda_2$ implies $E(\lambda_1) \leq E(\lambda_2)$.

ii) For $\lambda \to -\infty$ and $\lambda \to +\infty$, $E(\lambda)$ converges ultraweakly to 0 and 1, respectively. (For a "bounded" observable, with measured values confined to a finite interval $[\Lambda_1, \Lambda_2]$, we simply have $E(\lambda) = 0$ for $\lambda < \Lambda_1$ and $E(\lambda) = 1$ for $\lambda \geq \Lambda_2$.)

iii) For sequences $\lambda_i \ddagger \lambda$ converging to λ from above, $E(\lambda_i)$ converges ultraweakly to $E(\lambda)$, whereas for sequences $\lambda_i \ddagger \lambda$ converging to λ from below, $E(\lambda_i)$ also has an ultraweak limit, which is $\leq E(\lambda)$ (by i)) but need not be equal to $E(\lambda)$.

According to the famous spectral theorem for self-adjoint operators, every such spectral family $E(\lambda)$ uniquely determines a self-adjoint operator

$$A = \int \lambda \, dE(\lambda) \; ; \tag{6.63}$$

conversely, every self-adjoint operator A may be represented in the form (6.63) with a unique spectral family $E(\lambda)$ satisfying conditions i) to iii) above. (Conditions ii) and iii) are usually formulated by requiring strong rather than ultraweak convergence, but this makes no difference. We will not present here a rigorous version of the spectral theorem including, e.g., a rigorous definition of the integral in Eq. (6.63) (cf., e.g., [13], Ch. 4). - If the observable considered is bounded, (6.63) defines a bounded self-adjoint operator A. Such operators have been called Hermitean in the preceding chapters.)

Usually a quantum mechanical (decision) observable is represented by the operator A as given by (6.63), and is simply called "the observable A". We need not rederive here the well-known features of this description such as, e.g., the connection between the possible measured values of the observable and the spectrum of A, or the formulae

$$\langle A \rangle_W = \mathrm{tr}(AW) \tag{6.64}$$

and

$$(\Delta_W A)^2 = tr(A^2 W) - (tr(AW))^2 \qquad (6.65)$$

for the expectation value $<A>_W$ and the mean square deviation $\Delta_W A$ of an observable A in a state W; they follow immediately from the definition of A in terms of the spectral measure $\mathbf{E}$ and the physical meaning of the latter. (Compare also Eqs. (6.68) and (6.69) below.) If the observable considered - or, equivalently, the corresponding operator A - is unbounded, Eqs. (6.64) and (6.65) make sense for states W in a suitable "domain" only. For a discussion of this point see [14].

The spectral family $E(\lambda)$ is uniquely determined by the operator A, and the spectral measure $\mathbf{E} : \alpha \to E_\alpha$ may be reconstructed from $E(\lambda)$ - i.e., from A - because, in terms of the characteristic function

$$\chi_\alpha(\lambda) = \begin{cases} 1 & \text{for } \lambda \in \alpha \\ 0 & \text{otherwise} \end{cases}$$

of the Borel set α, E_α is given by

$$E_\alpha = \int \chi_\alpha(\lambda)\, dE(\lambda) = \chi_\alpha(A) \ .$$

The description of a decision observable by a self-adjoint operator A is therefore completely equivalent to the description by a spectral measure $\mathbf{E}$, and it is also practically useful since physically relevant quantities can be calculated directly in terms of A.

The preceding construction can be generalized to an arbitrary observable, as decribed by a POV measure $\mathbf{F} : \alpha \to F_\alpha$, and therefore simply called "the observable $\mathbf{F}$" in the following. In analogy to (6.62), one may define a "generalized spectral family"

$$F(\lambda) = F_{(-\infty, \lambda]} \ , \qquad (6.66)$$

with properties completely analogous to i) - iii) above. From $F(\lambda)$, the expectation value $<\mathbf{F}>_W$ and the mean square deviation $\Delta_W \mathbf{F}$ of the observable $\mathbf{F}$ in a given state W may then be calculated as Stieltjes integrals with the weight function

$$\omega(\lambda) = \mathrm{tr}(F(\lambda)W) \qquad (6.67)$$

according to the formulae

$$\langle \mathbf{F} \rangle_W = \int \lambda \, d\omega(\lambda) \qquad (6.68)$$

and

$$(\Delta_W \mathbf{F})^2 = \int (\lambda - \langle \mathbf{F} \rangle_W)^2 \, d\omega(\lambda) = \int \lambda^2 \, d\omega(\lambda) - (\langle \mathbf{F} \rangle_W)^2 \, . \qquad (6.69)$$

These formulae follow easily from the probability interpretation of

$$d\omega(\lambda) = \omega(\lambda+d\lambda) - \omega(\lambda) = \mathrm{tr}\big((F(\lambda+d\lambda)-F(\lambda))W\big) = \mathrm{tr}(F_{(\lambda,\lambda+d\lambda]}W) \, .$$

In analogy to (6.63), one may now attempt to define an operator

$$A = \int \lambda \, dF(\lambda) \, , \qquad (6.70)$$

in terms of which - at least formally - the expection values (6.68) could then be rewritten in the more familiar form tr(AW) (cf. (6.64)). If the observable $\mathbf{F}$ is bounded, (6.70) may indeed be shown - when interpreted appropriately - to define a bounded self-adjoint (i.e., Hermitean) operator A, such that $\langle \mathbf{F} \rangle_W = \mathrm{tr}(AW)$ for all W. But even in this simple case the right hand side of (6.69) can not be rewritten in the usual form $\mathrm{tr}(A^2W) - (\mathrm{tr}(AW))^2$ (cf. (6.65)), so that for a general observable $\mathbf{F}$ it is not possible to calculate mean square deviations via (6.65) directly from the corresponding operator A. Indeed, whereas for a decision observable (6.63) also implies $\int \lambda^2 dE(\lambda) = A^2$, and thus $\int \lambda^2 d(\mathrm{tr}(E(\lambda)W)) = \mathrm{tr}(A^2W)$, Eq. (6.70) does not imply $\int \lambda^2 dF(\lambda) = A^2$, so that $\int \lambda^2 d\omega(\lambda)$ need not be equal to $\mathrm{tr}(A^2W)$. Moreover, neither the POV measure $\mathbf{F} : \alpha \to F_\alpha$ nor even the generalized spectral family $F(\lambda)$ can be reconstructed uniquely from the operator A defined by (6.70), since a given operator A may have several different representations of the form (6.70). (For instance, the Hermitean operator A associated with a bounded observable $\mathbf{F}$ has, besides

(6.70), at least one additional representation of this form, namely the one provided by the spectral theorem, with the "ordinary" spectral family $E(\lambda)$ of A in place of $F(\lambda)$.) If the observable considered is unbounded, the definition (6.70) of the operator A leads to "domain problems" similar to, but even more severe than, the known ones associated with Eq. (6.63); actually the operator A need not even be densely defined, and in extreme cases its domain of definition may consist of the zero vector only. (Observables $\mathbf{F}$ leading to such difficulties are, however, rather "pathological" from the physical point of view as well.)

But even if - as for a bounded obervable - the operator A is well-defined, it provides a rather incomplete description of the corresponding observable, since not even mean square deviations can be calculated from it, and therefore it is much less useful than the self-adjoint operator associated with a decision observable. A general observable, therefore, has to be described either directly by the corresponding POV measure $\mathbf{F}$, or alternatively by the generalized spectral family $F(\lambda)$ associated with it, from which quantities of physical interest like $\langle\mathbf{F}\rangle_W$ and $\Delta_W\mathbf{F}$ can also be calculated via Eqs. (6.67) to (6.69).

Although quite natural in the version of quantum mechanics presented here, the consideration of such more general observables would still be merely of academic interest if it could not be illustrated by concrete physical examples. Such an application - in fact the only really interesting one known up to now - concerns the localization of massless particles. Since this subject has been discussed in much detail elsewhere [15], we shall sketch here the basic facts only.

A position observable, describing the spatial localization of a quantum mechanical particle at a fixed time, $t = 0$ say, is of a slightly more general type than the observables considered so far, since its "measured values" are the three spatial coordinates of the particle - i.e., triples of real numbers, rather than single ones. Therefore a position observable is to be described by a POV measure $\mathbf{F} : \alpha \to F_\alpha$ defined on the Borel sets α in three-dimensional space $\mathbb{R}^3$, rather than on the real line $I = \mathbb{R}^1$. An effect apparatus f_α measuring the effect F_α is realized physically - for finite and sufficiently simple α at least - by a particle counter occupying the spatial volume α and "operating" at time $t = 0$. (The last condition again represents an idealization, since a real counter is sensitive in some time interval, rather than at a sharp time only, and is thus expected to measure F_α only approximately at best.)

An elementary particle in relativistic quantum mechanics is characterized by its transformation properties under Poincaré (i.e., inhomogenous Lorentz) transformations, as described by a suitable irreducible unitary representation of the Poincaré group P on its state space H. This leads to an additional condition for the position observables of such particles: besides the conditions analogous to (6.52) and (6.54), the corresponding POV measure has to satisfy a certain covariance condition with respect to the given unitary representation of the Euclidean subgroup of P. Such POV measures are called Euclidean covariant in [15].

Conventional quantum mechanics admits only decision observables. Accordingly, particle counters are to be described by decision effects E_α, and the corresponding mapping $\mathbf{E} : \alpha \to E_\alpha$ becomes an Euclidean covariant spectral measure on $\mathbb{R}^3$. From such a spectral measure, a position operator

$$\mathbf{X} = \begin{pmatrix} X_1 \\ X_2 \\ X_3 \end{pmatrix}$$

with three mutually commuting self-adjoint components X_i can be constructed, and vice versa. It has been shown long ago by Newton and Wigner [16] - and later on, more rigorously, by Wightman [17] - that the above-mentioned requirements lead to a unique "Newton-Wigner" position operator $\mathbf{X}$ for massive particles and for massless particles with spin zero, whereas for massless particles with spin they can not be satisfied at all.

Massless particles with spin - e.g., photons or neutrinos - thus appear to be not localizable according to conventional quantum mechanics, whereas in practice at least photons can certainly be localized by suitable counters. This apparent discrepancy can indeed be resolved by admitting Euclidean covariant POV measures, rather than spectral measures only, as describing position observables. Thereby one arrives at a quite satisfactory description of position measurements for elementary particles of arbitrary mass and spin. For more details the interested reader is referred to the literature ([15], [7]).

References

[1] Kraus, K.: Operations and Effects in the Hilbert Space Formulation of Quantum Theory.
In: Hartkämper, A., Neumann, H. (Eds.): Foundations of Quantum Mechanics and Ordered Linear Spaces. (Lecture Notes in Physics, Vol 29.)
Springer, Berlin / Heidelberg / New York, 1974; p. 206 - 229.

[2] For the most recent account of Ludwig's theory, and for many additional references, see:
Ludwig, G.: Foundations of Quantum Mechanics I.
Springer, New York / Heidelberg / Berlin, 1983.

[3] Davies, E.B.: Quantum Theory of Open Systems.
Academic Press, London / New York / San Francisco, 1976.

[4] Schatten, R.: Norm Ideals of Completely Continuous Operators.
Springer, Berlin / Göttingen / Heidelberg, 1960.

[5] Haag, R., and Kastler, D., J. Math. Phys. 5, 848 - 861, 1964.

[6] Stinespring, W.F., Proc. Amer. Math. Soc. 6, 211 - 216, 1955.

[7] Kraus, K., Ann. Phys. (N.Y.) 64, 311 - 335, 1971.

[8] Gorini, V., Kossakowski, A., and Sudarshan, E.C.G., J. Math. Phys. 17, 821 - 825, 1976.

[9] Lindblad, G., Commun. Math. Phys. 48, 119 - 130, 1976; 65, 281 - 294, 1979.

[10] Naimark, M.A.: Normed Algebras.
Wolters-Noordhoff, Groningen, 1972; §22, No. 3.

[11] Wigner, E.P.: Group Theory and its Application to the Quantum Mechanics of Atomic Spectra.
Academic Press, New York / London, 1959; Ch. 26.

[12] Hellwig, K.-E., and Kraus, K., Commun. Math. Phys. 11, 214 - 220, 1969; 16, 142 - 147, 1970.

[13] Jauch, J.M.: Foundations of Quantum Mechanics.
Addison-Wesley, Reading (Mass.), 1968.

[14] Kraus, K., and Schröter, J., Intern. J. Theor. Phys. 7, 431 - 442, 1973.

[15] Kraus, K.: Position Observables of the Photon.
In: Pryce, W.C., Chissick, S.S. (Eds.): The Uncertainty Principle and Foundations of Quantum Mechanics.
Wiley, London / New York / Sidney / Toronto, 1977; p. 293 - 320.

[16] Newton, T.D., and Wigner, E.P., Revs. Mod. Phys. 21, 400 - 406, 1949.

[17] Wightman, A.S., Revs. Mod. Phys. 34, 845 - 872, 1962.

B. d'Espagnat

In Search of Reality

A physicist's view

Translated from the French by A. Ehlers
1983. 3 figures. IX, 183 pages
ISBN 3-540-11399-1

Contents: Introduction. - From Democritus to Pythagoras. - Philosophy of Experience. - Nonseparability. - Unkind Artless Interludes. - Comments on Scientism. - Einstein's Objections to the Philsophy of Experience. - Other Approaches: Elements for a Skepticism. - Veiled Reality. - Myths and Models. - Science and Philosophy. - Nonseparability and Contrafactuality. - Glances. - Conclusions. - Appendix I. - Appendix II. - Glossary. - References.

To what extent does contemporary physics make it possible to assert the existence or the nonexistence of an independent reality lying, at least in principle, within the comprehension of the human mind? Concerning that basic question, which has to be periodically reexamined taking the new accomplishments of science into account, the book brings to light fresh elements worthy of our attention and contributes a systematic reflection. Summarizing the results of recent developments in the field of particle physics as well as their consequences relevant to the problem, it compares modern physical theories with several philosophical approaches. Well-founded conclusions are drawn concerning both what can be expected from physics and what physics presumably will never be able to unveil. More generally, the book clarifies conclusively which philosophical approaches are compatible with these new results. Surprisingly enough, this investigation proves to have an important bearing on fields seemingly quite far apart from physics and on problems of considerable current interest. The book is written for a general audience interested in fundamental problems of science. It expresses its ideas in everyday language and avoid mathematical deductions which despite its logical accuracy, might be hardly intelligible for the majority of the readers. The author's original and far-reaching ideas will certain have an inpact on future discussions on the foundations of science and our perception of the world.

B. d'Espagnat

Auf der Suche nach dem Wirklichen

Aus der Sicht eines Physikers

Übersetzt aus dem Französischen von A. Ehlers
1983. 3 Abbildungen. IX, 191 Seiten
ISBN 3-540-12058-0

Inhaltsübersicht: Einleitung. - Von Demokrit zu Pythagoras. - Die Philosophie der Erfahrung. - Die Untrennbarkeit. - Böswillige und anspruchslose Zwischenspiele. - Bemerkungen über den Scientismus. - Die Einwände Einsteins gegen die Philosophie der Erfahrung. - Andere Möglichkeiten: Anregungen zur Skepsis. - Verschleierte Wirklichkeit. - Mythen und Modelle. - Naturwissenschaft und Philosophie. - Untrennbarkeit und Widersinnigkeit. - Streiflichter. - Schlußbemerkungen. - Anhang I. - Anhang II. - Lexikon. - Bibliographie. - Sachverzeichnis.

Wenn der renommierte französische Physiker und Denker Bernard d'Espagnat in diesem Buch die Frage nach dem Wirklichen aufwirft, überschreitet er die traditionelle Grundlagenforschung der Physik und leistet einen originellen Beitrag zu diesem tiefgehenden philosophischen Problem. Zwar bleibt er der Physik verhaftet, wenn er in ihr zunächst den besten Ansatz für eine Antwort sieht, und er wurzelt auch in der modernen Naturwissenschaft, wenn er dem mathematisch-logischen Schließen den Vorzug vor der intuitiven Spekulation gibt, doch zielt er weit über das hinaus, was der Naturwissenschaftler üblicherweise an Antworten bereithält. Eine Analyse ihrer modellgebundenen Verfahren führt den Autor zu der These, daß gerade die Physik uns zwingt, zwischen der Wirklichkeit und den Phänomenen zu unterscheiden. Ein modellunabhängiger Beweis zeigt, daß, sofern Experimente richtig deutbar sind, das Sein untrennbar ist, also mit den beobachtbaren Objekten nicht gleichgesetzt werden darf, die intrinsische Wirklichkeit also nicht auf die unserer gewohnten Existenz entliehenen Begriffe bezogen werden sollte.
Das Buch setzt keine Vertrautheit mit dem mathematisch-physikalischen Formalismus oder mit Einzelheiten aus Theorie und Experiment voraus. Vielmehr war es die Absicht des Verfassers, es auch für nicht-Naturwissenschaftler lesbar zu machen und diesen zu demonstrieren, wie die sich in den exakten Naturwissenschaften in den letzten hundert Jahren so explosiv entfalteten Denkformen neue Antworten auf alte und tiefe Fragen der Philosophie geben können.

Springer-Verlag Berlin Heidelberg New York Tokyo

Lecture Notes in Physics

Vol. 152: Physics of Narrow Gap Semiconductors. Proceedings, 1981. Edited by E. Gornik, H. Heinrich and L. Palmetshofer. XIII, 485 pages. 1982.

Vol. 153: Mathematical Problems in Theoretical Physics. Proceedings, 1981. Edited by R. Schrader, R. Seiler, and D.A. Uhlenbrock. XII, 429 pages. 1982.

Vol. 154: Macroscopic Properties of Disordered Media. Proceedings, 1981. Edited by R. Burridge, S. Childress, and G. Papanicolaou. VII, 307 pages. 1982.

Vol. 155: Quantum Optics. Proceedings, 1981. Edited by C.A. Engelbrecht. VIII, 329 pages. 1982.

Vol. 156: Resonances in Heavy Ion Reactions. Proceedings, 1981. Edited by K.A. Eberhard. XII, 448 pages. 1982.

Vol. 157: P. Niyogi, Integral Equation Method in Transonic Flow. XI, 189 pages. 1982.

Vol. 158: Dynamics of Nuclear Fission and Related Collective Phenomena. Proceedings, 1981. Edited by P. David, T. Mayer-Kuckuk, and A. van der Woude. X, 462 pages. 1982.

Vol. 159: E. Seiler, Gauge Theories as a Problem of Constructive Quantum Field Theory and Statistical Mechanics. V, 192 pages. 1982.

Vol. 160: Unified Theories of Elementary Particles. Critical Assessment and Prospects. Proceedings, 1981. Edited by P. Breitenlohner and H.P. Dürr. VI, 217 pages. 1982.

Vol. 161: Interacting Bosons in Nuclei. Proceedings, 1981. Edited by J.S. Dehesa, J.M.G. Gomez, and J. Ros. V, 209 pages. 1982.

Vol. 162: Relativistic Action at a Distance: Classical and Quantum Aspects. Proceedings, 1981. Edited by J. Llosa. X, 263 pages. 1982.

Vol. 163: J.S. Darrozes, C. Francois, Mécanique des Fluides Incompressibles. XIX, 459 pages. 1982.

Vol. 164: Stability of Thermodynamic Systems. Proceedings, 1981. Edited by J. Casas-Vázquez and G. Lebon. VII, 321 pages. 1982.

Vol. 165: N. Mukunda, H. van Dam, L.C. Biedenharn, Relativistic Models of Extended Hadrons Obeying a Mass-Spin Trajectory Constraint. Edited by A. Böhm and J.D. Dollard. VI, 163 pages. 1982.

Vol. 166: Computer Simulation of Solids. Edited by C.R.A. Catlow and W.C. Mackrodt. XII, 320 pages. 1982.

Vol. 167: G. Fieck, Symmetry of Polycentric Systems. VI, 137 pages, 1982.

Vol. 168: Heavy-Ion Collisions. Proceedings, 1982. Edited by G. Madurga and M. Lozano. VI, 429 pages. 1982.

Vol. 169: K. Sundermeyer, Constrained Dynamics. IV, 318 pages. 1982.

Vol. 170: Eighth International Conference on Numerical Methods in Fluid Dynamics. Proceedings, 1982. Edited by E. Krause. X, 569 pages. 1982.

Vol. 171: Time-Dependent Hartree-Fock and Beyond. Proceedings, 1982. Edited by K. Goeke and P.-G. Reinhard. VIII, 426 pages. 1982.

Vol. 172: Ionic Liquids, Molten Salts and Polyelectrolytes. Proceedings, 1982. Edited by K.-H. Bennemann, F. Brouers, and D. Quitmann. VII, 253 pages. 1982.

Vol. 173: Stochastic Processes in Quantum Theory and Statistical Physics. Proceedings, 1981. Edited by S. Albeverio, Ph. Combe, and M. Sirugue-Collin. VIII, 337 pages. 1982.

Vol. 174: A. Kadić, D.G.B. Edelen, A Gauge Theory of Dislocations and Disclinations. VII, 290 pages. 1983.

Vol. 175: Defect Complexes in Semiconductor Structures. Proceedings, 1982. Edited by J. Giber, F. Beleznay, J. C. Szép, and J. László. VI, 308 pages. 1983.

Vol. 176: Gauge Theory and Gravitation. Proceedings, 1982. Edited by K. Kikkawa, N. Nakanishi, and H. Nariai. X, 316 pages. 1983.

Vol. 177: Application of High Magnetic Fields in Semiconductor Physics. Proceedings, 1982. Edited by G. Landwehr. XII, 552 pages. 1983.

Vol. 178: Detectors in Heavy-Ion Reactions. Proceedings, 1982. Edited by W. von Oertzen. VIII, 258 pages. 1983.

Vol. 179: Dynamical Systems and Chaos. Proceedings, 1982. Edited by L. Garrido. XIV, 298 pages. 1983.

Vol. 180: Group Theoretical Methods in Physics. Proceedings, 1982. Edited by M. Serdaroğlu and E. Inönü. XI, 569 pages. 1983.

Vol. 181: Gauge Theories of the Eighties. Proceedings, 1982. Edited by R. Raitio and J. Lindfors. V, 644 pages. 1983.

Vol. 182: Laser Physics. Proceedings, 1983. Edited by J. D. Harvey and D. F. Walls. V, 263 pages. 1983.

Vol. 183: J. D. Gunton, M. Droz, Introduction to the Theory of Metastable and Unstable States. VI, 140 pages. 1983.

Vol. 184: Stochastic Processes – Formalism and Applications. Proceedings, 1982. Edited by G. S. Agarwal and S. Dattagupta. VI, 324 pages. 1983.

Vol. 185: H. N. Shirer, R. Wells, Mathematical Structure of the Singularities at the Transitions between Steady States in Hydrodynamic Systems. XI, 276 pages. 1983.

Vol. 186: Critical Phenomena. Proceedings, 1982. Edited by F.J.W. Hahne. VII, 353 pages. 1983.

Vol. 187: Density Functional Theory. Edited by J. Keller and J.L. Gázquez. V, 301 pages. 1983.

Vol. 188: A.P. Balachandran, G. Marmo, B.-S. Skagerstam, A. Stern, Gauge Symmetries and Fibre Bundles. IV, 140 pages. 1983.

Vol. 189: Nonlinear Phenomena. Proceedings, 1982. Edited by K. B. Wolf. XII, 453 pages. 1983.

Vol. 190: K. Kraus, States, Effects, and Operations. Edited by A. Böhm, J. W. Dollard and W. H. Wootters. IX, 151 pages. 1983.

Selected Issues from

Lecture Notes in Mathematics

GPSR Compliance

The European Union's (EU) General Product Safety Regulation (GPSR) is a set of rules that requires consumer products to be safe and our obligations to ensure this.

If you have any concerns about our products, you can contact us on ProductSafety@springernature.com

In case Publisher is established outside the EU, the EU authorized representative is:

Springer Nature Customer Service Center GmbH
Europaplatz 3
69115 Heidelberg, Germany

Batch number: 10260223

Printed by Printforce, the Netherlands